AF413193

Chief Impact Officer

Chief Impact Officer

*Real transformation comes from human—
not just artificial—intelligence.*

< JULIE AVERILL

**former CIO of lululemon and REI,
Nordstrom exec**

an imprint of Microsoft

© 2026 Microsoft Corporation. All rights reserved.

Distributed and printed by Simon & Schuster

Published in the United States by 8080 Books,
an imprint of Microsoft Corporation

1 Microsoft Way, Redmond, WA 98052

https://aka.ms/8080books

Cover Design by Shyla Lindsey
Typesetting by Nord Compo
In association with Wise Wolf Creative

ISBN 979-8-9937553-8-0 (Paperback – 8080 Books)
ISBN 979-8-9937553-9-7 (Hardcover – 8080 Books)
ISBN 979-8-9937554-4-1 (eBook – 8080 Books)

MICROSOFT, 8080 BOOKS, and the 8080 BOOKS Logo
are trademarks of Microsoft Corporation

Praise for *Chief Impact Officer*

The Imperative of People and Culture in a Tech-Driven World

In today's rapidly evolving, technology-centric landscape, it's easy for C-level executives to become singularly focused on digital transformation and AI integration.

However, I firmly believe that the true competitive differentiator isn't found solely in our code or our cloud infrastructure, but in the people who build and maintain them.

We must treat our organizational culture not as a soft, secondary metric, but as a strategic asset. A vibrant, inclusive culture that encourages psychological safety, continuous learning, and experimentation is the foundation upon which innovation is built.

Without a strong culture, even the most sophisticated technology stack will fail to achieve its potential, leading to burnout, high turnover, and a stagnation of creative problem-solving. Prioritizing culture ensures we retain top talent and cultivate an environment where they are motivated to bring their best ideas to the table every single day.

The second critical component is the commitment to growing every single employee to their fullest potential. In a world where the half-life of a technical skill is constantly shrinking, talent development is no longer a luxury—it's an operational necessity.

We must invest heavily in personalized learning paths, mentorship, and cross-functional opportunities that challenge our people to acquire skills beyond their immediate job description. This commitment not only future-proofs our workforce against technological obsolescence but also fosters a deep sense of loyalty and purpose within the organization. When employees see a clear path for growth, they are more engaged, more resilient to change, and far more likely to contribute to long-term corporate success. It's a simple equation: we grow our people, and in turn, they grow the business.

Ultimately, the most successful enterprises in the next decade will be those that strike a harmonious balance between cutting-edge technology and human potential. As C-level leaders, our mandate is to be the architects of this balance. This means systematically dismantling silos, rewarding collaboration over individual heroism, and embedding our core values into every decision we make, from hiring to product development.

By fostering a culture of continuous improvement and empowering our people with the resources and autonomy they need, we ensure that our people remain the most adaptable and powerful engine of innovation we possess. This isn't just a people strategy; it's the definitive strategy for sustainable, market-leading growth in the modern era.

This book provides a comprehensive roadmap from a seasoned industry leader who fully understands how to make this mandate for sustainable excellence in an ever-changing world a reality.

—Mike Richardson,
retired, former CIO Nordstrom

Table of Contents

Foreword

When you've spent your career at the intersection of technology and trust, you learn a simple truth: transformation isn't about tools—it's about people. I've had the privilege of leading through major technology waves over the past three decades: the rise of the web, the explosion of mobile, the migration to cloud, and now the era of AI. Each wave promised speed, scale, and efficiency. Each delivered those things—But only for leaders who understand that culture is the most important differentiator in transformation. Even the best strategy and transformation plan will fail if the culture required to execute it is weak, misaligned, or toxic. A brilliant plan doesn't matter if the people, behaviors, norms, and values of the organization don't support it.

Strong foundations accelerate progress. Weak ones collapse under pressure.

That's why this book matters.

I've known Julie Averill for more than ten years, she doesn't just modernize systems—she rewires organizations for resilience. She doesn't just deploy technology—she builds cultures where people can speak truthfully, take risks, and own outcomes. In a world obsessed with disruption, Julie reminds us that the most powerful force for change isn't an algorithm. It's a leader who knows how to bring people along with them. People don't execute a strategy because a slide tells them to—they execute when they are engaged, supported, and aligned.

Julie's career spans some of the most iconic brands in retail—Nordstrom, REI, and lululemon—where she turned technology from a back-office function into a strategic capability. But what sets her apart isn't the scale of those transformations. It's the way she led them: with clarity, courage, and a steadfast belief that people are the multiplier, not the constraint. When lululemon's website crashed three days before she officially started, she didn't have authority yet. What she had was influence—the ability to rally a team around a shared goal and turn crisis into opportunity and the seed for cultural change. That moment became the foundation for an eight-year journey that scaled the company from $2 billion to over $10 billion in revenue.

I've seen the same pattern play out across industries. The leaders who thrive in times of change aren't the ones with the biggest budgets or the fanciest tech stacks. They are the ones who can create trust in a way that fosters expertise to surface early. Who can unite teams across boundaries. Who can make progress through uncertainty and develop others so capability scales. These aren't soft skills. They are performance skills in a world where change never waits. This is how successful leaders create a culture that enables their teams and people to reach their full accretive value.

AI makes this more urgent than ever. It shifts power toward the people who understand the data scientist who sees a pattern no executive can, a product manager who knows what collaboration makes possible, a store manager who spots friction before headquarters does.

AI makes this more urgent than ever. It shifts power to the people who understand the details, ask better questions and drive decisions closer to actual work. Hierarchies slow you down. Influence moves you forward. That's why the capabilities Julie teaches in this book—power fluency, sensemaking, psychological safety, human grounding, and inclusive capacity-building—aren't optional. They're the operating system for leadership in the AI era.

What I love about this book is that it's not theory. It's lived experience. Julie takes you inside the moments that shaped her leadership: the childhood lessons in systems thinking from a catcher's crouch, the quiet strength learned from a mother who refused to accept limits, the hard truths about authenticity that came from years of hiding who she was. She shares the crises that tested her—outages, acquisitions, pandemics—and the decisions that defined her: consolidating teams to build trust, killing failed projects openly, embedding innovation into operations instead of isolating it. These stories aren't just compelling. They're practical. Each chapter ends with actions you can take now, because transformation doesn't wait for perfect conditions. It starts with what you do next.

Security, Trust, and Influence

In my experience leading digital transformation at Microsoft, especially during my tenure as Chief Information Security Officer, trust was the foundation of everything we did

Security isn't just about technology or defense—it's about confidence. It's the foundation that lets you move faster, recover quickly, and innovate safely. It shouldn't slow progress, it should make progress possible. People need to believe that their data, their privacy, and their future are protected. That belief doesn't come from control alone. It comes from influence.

I often say: **"Sphere of influence is always more important than span of control."** In security, you rarely own every system, every process, or every decision. You succeed by influencing people across boundaries—helping them understand risks, make better choices, and embed security into the way they work. That's true for every transformation. Authority can enforce compliance. Influence creates commitment. And commitment is what scales.

Julie understands this better than anyone. Her approach to leadership mirrors what I've seen in the most resilient security programs: build trust early, distribute responsibility, and create environments where people surface problems before they become crises. Whether you're securing a global enterprise or scaling AI responsibly, the principle is the same: influence moves faster than control.

And in a world where technology changes daily, speed matters.

A Personal Perspective

One of my favorite memories of working with Julie came during a cross-industry security and technology leadership forum. We were discussing the challenge of embedding security into innovation without slowing down progress—a tension every leader feels. Julie didn't just offer a solution; she reframed the problem. She said, "f security feels like a gate, you've already lost. It must feel like a guardrail—something that enables speed, not limits it." That insight sparked a conversation that changed how several companies approached secure innovation. It was classic Julie: clear, practical, and deeply human. She didn't dominate the room with authority; she influenced it with wisdom. That's leadership in action.

I've spent my career helping organizations protect what matters most—security, privacy, trust—while enabling innovation at scale. I've seen what happens when leaders treat transformation as a technology problem instead of a human one. They buy tools instead of building capability. They chase pilots instead of defining success. They create complexity that sends the bill later, with interest. Julie's approach is the antidote. She shows you how to start with clarity, build the muscle before you add the weight, and lead in a way that makes people believe in the future you're asking them to create.

If you're holding this book, you're probably facing a moment of change. Maybe you're implementing AI. Maybe you're scaling globally. Maybe you're trying to turn a vision into reality without all the answers. Whatever your context, here's my advice: read this book thoughtfully. Sit with the questions Julie asks. Apply the principles she shares. Because the future isn't just about technology. It's about trust. And trust is built one decision, one conversation, one act of courage at a time.

Julie Averill has given us more than a playbook for transformation. She's given us a blueprint for leadership that lasts. In a world where algorithms will keep getting smarter, this book reminds us of something essential: the most powerful intelligence in any organization is still human.

—Bret Arsenault
Chief Information Security Officer: Emeritus Microsoft

Introduction

Transformation with Impact

Leaders everywhere are being asked to deliver AI-driven change. New tools, new budgets, big expectations. But what will set successful leaders apart isn't how much technology they deploy. It's how well they develop the human capabilities that turn tools into impact.

You can approve funding. You can set priorities. But results only happen when people trust the direction, feel ownership in the work, and can speak up when the truth is uncomfortable. That's been at the center of everything I've learned as a leader, and it's the thesis of this book: **real transformation comes from human, not artificial, intelligence.**

AI makes human capability more important, not less. It changes how decisions get made and who has insight worth hearing. It rewards leaders who build alignment early, learn fast, and create space for others to contribute.

The Pattern We've Seen Before

Every major technology wave creates urgency. And urgency leads to shortcuts.

During the dot-com years, companies rushed to build websites because everyone else had one. The winners asked what the internet could make possible for their customers. Cloud adoption followed the same pattern. Many expected automatic savings and speed. The leaders who gained the most understood that cloud wasn't a destination—it was a different way of operating.

The same thing is happening with AI. Some treat it as a project or a purchase. Others see it as an opportunity to rethink how they create value.

The difference comes down to leadership.

Why this Matters Now

When you've already built trust, fast decisions become possible. I learned that at Nordstrom. We made a call about inventory visibility that created a $250 million revenue lift within months. That wasn't magic. It was the result of relationships, clarity, and accountability that existed before we needed them.

At lululemon, I learned the opposite lesson. Two days before I officially started, the website went down for twenty hours. I discovered the outage, but more importantly, I found the systems underneath it—the people, the processes, the technology. That moment showed me ownership, transparency, and a bias for action was what we needed to build.

What both lessons taught me is that transformation is a human problem first. Technology just amplifies whatever capability already exists.

And right now, AI is forcing that truth into the open.

You can't bolt artificial intelligence onto broken processes and dysfunctional teams and expect transformation.

The technology exposes unclear decision rights, territorial behavior, and people hoarding information instead of sharing it. AI doesn't fix your culture. It reveals it.

At the same time, the workforce has changed. Gen Z watched their parents perform loyalty to companies that laid them off by spreadsheet. They're not buying it. They want to be part of something real, and they'll walk away from prestige and paychecks if the culture is hollow or the leadership is phony. They're right to.

This convergence is what makes the old playbook fail. Command and control, information as power, transformation by decree—none of it works anymore. Not for harnessing AI's potential, and not for earning commitment from people who have options.

The leaders who will make it are the ones who build trust before they need it, who tell the truth even when it's uncomfortable, and who understand that technology is the easy part.

This book is for anyone leading change without absolute authority. Which, let's be honest, is most of us.

You might be on a board trying to figure out which human capabilities actually matter for AI strategy. You might be an executive who knows your technology investments will fail without the right culture. A product leader who can't ship without cross-functional commitment. An HR leader tired of culture being dismissed as "soft stuff." A manager translating strategy into something people can execute. An entrepreneur with nothing but influence.

Or you might just want to have real impact in your work. This book will change how you see leadership—not as a position, but as a choice you make about how you show up. If your impact depends on other people believing in the work, this book gives you the tools.

What You Will Learn

You'll see real examples from Nordstrom, REI, and lululemon that show how transformation works when technology and people advance together.

Part 1, "Foundations," explores what shaped me before I held a leadership title. My father teaching me systems thinking from a catcher's crouch. My mother showing me that strength doesn't announce itself. The cost of hiding who I was. Early learnings at Nordstrom and REI where I learned how influence really works.

Part 2, "Transformation," takes you inside the lululemon years. The opportunities seized, the foundations built, and the culture that let us scale from $2 billion to over $10 billion. Building leadership before scaling systems. Maintaining high standards while creating safety. Expanding globally while staying local. The pandemic that proved everything we'd built was real.

Part 3, "Leading Forward," examines what endures when you're gone. When to stay and when to leave. What it means to bet on yourself. And the question every leader eventually faces: was the transformation real, or did it depend on you being there to push it?

Throughout, you'll find sections that connect the stories to capabilities you can develop now. Not because AI is the only transformation that matters, but because it's accelerating every other one.

The Invitation

Technology doesn't transform companies. People do. AI will amplify whatever leadership exists, strong or weak.

The goal isn't to build better workers. It's to develop better humans who happen to do extraordinary work because you helped them become more capable, more confident, more fully themselves.

That's what this book is about. Let's begin.

Part 1:
Foundations

The capabilities that matter most for transformation take time to build. And sometimes you don't know you're building them until you look back and see that they became the foundation upon which you built everything.

Systems thinking. Self-worth that doesn't depend on external validation. The courage to speak truth when silence is easier. The ability to build trust across difference. These aren't skills you can acquire in a weekend workshop or from a consulting framework. They're patterns formed through experience, tested under pressure, and refined over time.

The chapters in this section aren't chronological career milestones. They're the foundational experiences that shaped how I think, lead, and transform. Some came from baseball fields and family businesses. Some came from corporate environments that taught me what I wouldn't accept. Some came from adopting a son across continents and learning what real partnership looks like.

You're building your own foundation right now whether you realize it or not. The question is whether you're paying attention to what the experiences are teaching you.

welcome to lululemon
lululemon.com
404
Namaste.

CHAPTER 1

Before I Even Started

Own What's Broken

May 22, 2017—two days before I officially joined lulu-lemon—I was sitting in my living room with my wife Cindy and our oldest son Mason. They were watching the NBA playoffs. I was half-listening to the game while on the phone with the lululemon interim CIO, getting oriented. "Check out the ABC pant," he said. "One of our bestsellers."

I pulled up the website, clicked on a product, and read "temporary savasana. We're usually awesome at this." It was a 404 Error. Another product. Another cute message.

"Hey," I said, "a couple of product pages aren't working."

He paused. "Which ones?"

"Shouldn't all of them be working?"

He asked me to screenshot and email it. I wanted to get the team on the phone and treat it as a high-priority incident, but I wasn't an employee yet. So, I sent the email.

That email became my introduction to the team. The site stayed down for eighteen more hours. The CEO blamed IBM publicly two weeks later. But the real problem wasn't the technology.

The site was fully outsourced. We had no visibility, no monitoring, no escalation path. Oracle confirmed the site was down but didn't know why. The company was on an old version of Oracle's ATG software and the support was limited.

I picked up the phone. I called my contacts at Oracle and learned the data center hosting the lululemon website had been sold to IBM and nobody knew who was responsible anymore. Then I called an executive at IBM who tracked down the data center owner. They didn't realize they were hosting lululemon.com.

Eventually, through a string of phone calls and referrals, I found the issue. An entire data center had gone down. It happened to be hosting lululemon's site.

Our site was down for *twenty hours*. Those twenty hours told me everything I needed to know. We weren't broken. There was just no intentional design, no architecture, no bigger purpose. That's where transformation starts.

No alerts. No drop detection. No ownership.

Those twenty hours felt endless. I hadn't even started and was already in crisis mode. My boss was panicking, my team was calm. There was nothing they could do. The vendors seemed indifferent.

I questioned my decision to join before I even walked through the door. I'd spent decades building expertise, reputation, credibility. I'd taken a role at a company I barely knew, drawn in by the brand's potential. Now I was sitting in my Seattle living room, watching my new company's digital presence disappear into the void.

All I knew was twenty hours of silence from a brand I'd just bet my career on.

Three days into my tenure, Jim Cramer mentioned it on CNBC, "Julie Averill better get to work."

He wasn't wrong.

That outage taught me what I should have already known, which is that you can't transform a company if you don't own what matters most. Real transformation starts with claiming control. We didn't need to build everything, but we needed to own it.

That brittleness wasn't just technical. It revealed something deeper. Leaders can't outsource ownership. Authentic leadership begins by facing what's broken and claiming responsibility for it.

My family and I were about to go on an eight-year rocket-ship ride on this unicorn of a company—helping grow it nearly 5x in eight years, earning a global reputation, and pushing every boundary to innovate, imagine what was possible, and create an unbelievable future. But, I'm getting ahead of myself.

You Can't Outsource What Matters Most

In my first month, as I was scrambling to get the team together and build a plan for the new website, I first needed to make sure the old one didn't collapse. I called Oracle, our hosting partner, to talk about website resilience. They were the same vendor behind the twenty-hour outage. I wanted to know what they planned to do about it.

Now, for perspective and fairness, Oracle had received very little direction from lululemon prior to this. The three confident salesmen (yes men), walked in polished and confident, schooled from years of an easy relationship. They brought multiple copies of a bound PowerPoint. I wanted a conversation, not a presentation, but took deep breathes to respect the history.

The lead consultant opened his laptop, smiled, and said, "We think you'll be pleased with this solution. As a new CIO, you want to have the peace of mind to know that if the site goes down again, it won't be 20 hours." It probably would have been better if they had started with an apology for all the revenue lost with the outage they were responsible for.

Those deep breathes turned to shallow ones rather quickly, as I flipped to the price tag for this "peace of mind." $6 million. Flowcharts. Diagrams. Slides promising "enterprise-grade redundancy." I could tell they expected easy approval. New CIO, big brand, fresh budget. My questions seemed to surprise him.

"Let me make sure I understand this," I said. My signature phrase for "this makes no sense and I'm going to show you why."

"You're not proposing any way to ensure the site stays up, such as monitoring or actual resilience. Instead you're proposing another data center with another instance of the website, that I would have the option to manually light up after being down for at least six hours. Is that correct?"

He didn't flinch. "Julie, this is the gold standard in the industry. The biggest retailers use it. What you're really buying is peace of mind."

"That's not prevention. That's waiting to fail," I said.

This reminded me of early in my career when a senior engineer told me, "Julie, let me put this in *simple terms* that you can understand." I have more, shall we say, executive presence now than I did at 22, thankfully. I just smiled and let him continue, knowing that I would speak my truth soon. I didn't need peace of mind. I needed accountability. And I knew that was something I needed to build, not buy.

"I appreciate the effort," I said. "But six hours of downtime isn't acceptable for a company growing this fast. We need proactive monitoring, real-time alerting, automatic recovery. If the site goes down, our guests shouldn't even notice."

The room went quiet. My infrastructure leads were listening as closely as the vendors. For years, they'd been trained to manage vendor relationships, not demand outcomes.

That meeting made one thing clear. We weren't going to rebuild our foundation by buying more decks or deferring decisions. We had to put our hands back on the wheel. But I also knew owning everything wouldn't fix anything if we didn't know what we were building toward. That's where vision stopped being a poster on the wall and became the thing that truly guided decisions.

Vision as a Filter

Our technology vision was never "let's move to the cloud" or "let's modernize our tech stack," though we did both of those things. It was about building a company that could scale without breaking, stay solid when things got tough, work globally, and still feel human. Technology was how we'd get there, not the point.

The company was growing at hyper scale, and the vision had to match that reality.

This vision became the filter for every decision. Which vendors to pick, which projects got funding, which teams to build first. When something didn't serve the vision, it was easier to say no.

The companies that win treat vision like a north star. They understand the true levers of the company and aim to find innovative ways to add value. Those that just spin their wheels treat vision like a tech roadmap with a to-do list. They chase pilots and fall behind while the board reads success in the quantity of "AI initiatives" underway. I'm watching it play out in real time today.

I sit in forums with AI founders who have real solutions, proven products. They keep asking the same questions, such as "How do I get the company to understand the impact?" or "IT wants to do a pilot, but the business side isn't involved." They're not struggling because their technology doesn't work. They're struggling because companies are treating AI like a technology problem instead of a business transformation problem.

A retail company piloted an AI allocation tool to crack merchandising's toughest problem of getting the right product, in the right size and color, in the right store, at exactly the moment a customer walks in. The AI analyzed nine thousand data points per store and created hyper-specific buy recommendations for each location.

Until the head merchant saw the first recommendations and asked, "How will this match up with our quarterly marketing campaigns?"

The AI was optimizing for sales. The business was optimizing for sales, brand consistency, marketing alignment, vendor relationships, margin targets, and seasonal storytelling. The pilot stalled right there.

The assignment coming out of that meeting was for merchants to document their decision-making process in a way that could be translated to an algorithm. That was two years ago. They're still working on it.

That's what happens when you pilot without strategy. You solve the wrong problem brilliantly.

Companies that link technology to clear vision move faster. They focus on what matters to customers and build technology that supports how decisions are truly made. Without that clarity, AI stalls. Not because the pilots are wrong, but because no one knows what success looks like.

Vision isn't a slogan. MIT recently released a study showing that 95 percent of all AI pilots fail. Not because the AI doesn't work. Because companies are adding intelligence to processes they don't fully understand[1].

Complexity Always Sends the Bill

From my earliest days writing code, I knew that complexity compounds into something dangerous. It's not sophistication. It's a red flag. And it's oh so difficult to clean up.

Every time I walked into a mess, the complexity had piled up slowly, one decision at a time that seemed totally reasonable in the moment. Nobody wakes up planning to build a disaster. But that's what happens when you optimize for today and forget about tomorrow.

I'd seen this pattern before. At a previous company, we'd been early adopters of cloud technology. For a while, that felt like an advantage. We could build faster than competitors. But then the bills showed up, and we realized that innovation without guardrails is just expensive chaos. Cloud costs had ballooned into millions we couldn't explain. Teams were spinning up whatever they needed without thinking about what it cost. We had to stop everything and spend months just figuring out what we even had.

The cleanup was brutal. Teams pushed back when they had to pay for their cloud hosting fees. A resource that was previously free was now a budget line item they had to manage. Leaders who'd been praised for moving fast were now getting asked to justify every line item. But the alternative was watching our infrastructure budget eat money we needed to actually grow.

That lesson was still fresh when I walked into lululemon.

At lululemon, I made the opposite mistake. I could see the complexity building, but I kept putting off the hard conversations about governance because we were moving so fast. Again, what was simple was not easy when we were moving so fast.

I attempted to slow the number of small projects down to leave room for the more strategic. To give me time to meet with business partners and get a view on what was important and what was not. It seemed like an easy feat to slow down. Every Monday was an "approval meeting" where project managers presented to the CIO their requests, mostly small and incremental. I cancelled the meetings and told the directors and our one VP that we would be building a process for approvals, and, in the meantime, they could have a conversation directly with me.

Yes, things quieted down. But all I had done was cut myself out of the loop. There were no controls. This fast-growing, innovative, "you can do anything" company would not slow down just because I came in with a title.

The 30-day assessment laid it all out. We had systems built without any real plan, vendor contracts signed under pressure, and teams drowning in technical debt. I called it "the architecture of fragmentation." Over two hundred projects running at once with almost no oversight, each one adding another thread to an already tangled web.

The vendor mess was particularly bad. For years, we'd been signing contracts to fix immediate problems without thinking about how it all fit together. Need analytics? Sign with Vendor A. Need security? Vendor B has something. Marketing wants automation? Sure, Vendor C can do that. Before long, we had hundreds of vendors, overlapping solutions, and zero leverage with any of them.

It all came to a head when a possible fraud incident forced us to audit the entire vendor landscape. What had felt like speed looked more like exposure. We had no central view of who had access to what data. Contracts were spread across different teams. Some vendors we'd honestly forgotten we were paying.

Fixing it meant letting people down and unwinding deals. It meant putting guardrails in place. But on the other side, we had fewer vendors, stronger relationships, and actual strategic partnerships, AKA leverage, for the first time. We could finally see the whole picture.

Simplicity is knowing what matters and letting everything else fall away. It's about clear boundaries and intentional connections. When you understand how things fit, you can move fast at the edges without breaking what matters. But when complexity runs wild, every change is a risk, and every new tool is another point of failure.

The hairball we created with fragmented systems took years to build. Today, companies are creating AI hairballs in months. They're locking themselves into contracts they can't untangle, tools that don't talk to each other, and solutions nobody uses because nobody chose them.

But I'm also seeing what works. Companies that start with a clear problem and a clear focus to solve it. They let their teams experiment, learn what actually helps, and then standardize around what works. They ask "what problem are we solving?" before they ask "what AI can we buy?"

These companies aren't moving slower. They're moving smarter. The goal isn't to be "AI-ready." It's to be clear about what you're trying to achieve and intentional about how AI helps you get there.

The winners will be the ones who resist the pressure to sign everything and instead build simple, solid foundations before they scale the complexity. Start with clarity. Add tools that serve it.

Build the muscle before you add the weight.

Putting off simplicity always feels cheaper right now. But complexity compounds, and the bill always shows up. These cleanups weren't just about technology. They were about building trust in the systems so people could lead without second-guessing everything. Simplicity became a leadership discipline.

That first month taught me that this wasn't a company that would sit around waiting for perfect information. The urgency was real. If I was going to lead transformation here, I needed to match that speed while building the discipline they didn't have yet.

What to Carry Forward

You Don't Need a Title to Lead. Two days before I officially started, I had no title, no team, no access. But the website was down. I picked up the phone, worked my network, and found the problem. You don't need authority to lead in a crisis. You need relationships, credibility, and the willingness to act. The leaders who wait for permission to lead rarely get it. The ones who step in when something's broken earn the authority that follows.

Own the Outcomes, Not Just the Tools. The lesson isn't "never outsource." It's "never outsource accountability."

You can partner with vendors for execution. But you must own the outcomes, own the relationships, and own the ability to see when things break.

Vision is a Filter, Not a Slogan. Our vision wasn't aspirational fluff. It's operational clarity. It's the answer to "what are we actually trying to achieve?" that lets you evaluate whether a tool, vendor, or initiative serves that goal. When you're clear about where you're going, every decision gets simpler.

Simplicity is Discipline. Complexity always sends the bill. It just doesn't arrive when you're making the decisions that create it.

Build the muscle before you add the weight.

Start with one clear problem. One solid solution. Build capability before scaling complexity. Ask "what problem are we solving?" before "what tech should we buy?"

These principles don't just make transformation faster. They make it sustainable. When your team owns outcomes, shares vision, and maintains simplicity, they can continue transforming after you leave. Authority-driven change stops when the leader stops pushing. Capability-driven change compounds.

These lessons didn't come from business school or consulting frameworks. They came from experiences that taught me to see systems and stand firm long before I had the language for it. My foundation was built earlier than I realized. It was built watching my dad read a baseball field and my mom refuse to accept limits others tried to place on her.

CHAPTER 2

Behind Home Plate

Los Angeles to Seattle

The leadership foundations that would carry me through billion-dollar transformations weren't learned in conference rooms. They were shaped long before anyone called me a leader. They came from a childhood where systems mattered, ownership wasn't optional, and strength looked quiet but unbreakable. Everything I later built as a CIO started here.

The foundation that would later hold through a twenty-hour website crash, a global pandemic, and the transformation of a multi-billion dollar company wasn't built in any meeting room I ever entered.

It was built in a house where my dad taught me to see the whole field from a catcher's crouch. Where my mom fixed her own thumb, twice, after accidentally sewing a needle through it and went back to work without complaint. Where strength looked nothing like what gets celebrated in corporate America, and leadership meant something entirely different from what I'd later encounter in my first management role.

These weren't leadership principles I studied. They were patterns I lived. And they would prove more valuable than any advanced degree when everything started breaking at once.

Learning Strategy from Behind Home Plate

My father, Earl Averill, had a personality that filled a room and a laugh that echoed down the hall. He spent seven years in baseball's Major Leagues, the son of a Hall of Famer, also named Earl Averill. If you're from

Figure 1 Earl Averill, Chicago Cubs 1960

Cleveland, perhaps you've seen the retired number 3 at Progressive Field. That was Grandpa. His number is the first one listed, right next to Bob Feller's #5. If you're from Anaheim, maybe you've seen pictures of the first Angels team. My dad was there. But you'd never hear Dad bring that up unless you asked. He didn't care much about titles or status. He cared about people. About relationships. About showing up.

What stayed with him wasn't just every play from his career (though he did remember every pitch and every play). It was every person. He had been a catcher, and that position shaped him profoundly. While the rest of the team faced the batter, he crouched low, looking out at both his team and the competitors. He saw the whole field. He noticed things. He paid attention. That vantage point of watching the game from behind, thinking two steps ahead, protecting and guiding, was more than a position. It was who he was.

Every time we'd go to a Mariners game together, I'd watch him observe and think. He knew the batters and their patterns, the pitchers and their tendencies, the way momentum shifted between innings. He would call the play before it happened, "Watch this. The pitcher is going to throw him outside and it's going over the right field fence." And that's exactly what happened.

He'd narrate the game, "See how the shortstop just shifted three steps toward second? He knows this batter pulls everything. Now watch the second baseman cheat toward the bag. They're setting up for the double play."

The count, the inning, the season standings, the scouting reports no one in the crowd could see. He trained his instincts by watching patterns and predicting outcomes.

Years later, when I walked into my first technology crisis, I instinctively moved to that catcher's position. Not rushing to fix the obvious problem in front of me, but stepping back to see the whole field. Understanding that the website outage wasn't really about technology; it was about fragmented teams, unclear ownership, and vendor relationships that had shifted power away from the people who should have held it.

Every time I've walked into a transformation situation, I've gone looking for what's really happening beneath the surface. Not just the failed system or the missed deadline, but the organizational dynamics that created those failures. The incentives that shaped behavior. The gaps between what people said mattered and what actually got rewarded.

Dad was one of those people who had a good friend named "Bob" and when you asked which Bob, the answer could be one of six. Bob the computer guy, Bob who owned the mortuary, Bob in Alaska. You name it. He built trust by being genuinely interested in people, and that trust became his greatest asset.

My friends adored him, especially my guy friends. One day I came home from a friends house to find eight of them hanging around the kitchen table with Dad. They had stopped by to see him and his model airplanes, baseball nostalgia, and fishing gear. But what I really saw was someone warm, curious, and fully alive in every interaction.

Dad and I talked or texted every day. Every single day. Even in his eighties. Even as I traveled. I video chatted with him from India to show him the bats flying overhead, from Ethiopia where I had just adopted his youngest grandson, and in Hong Kong where I was meeting my team for the first time. He loved knowing what was happening in my world. He devoured the details of my job. He didn't always understand the context, but that didn't matter. What mattered was that we were connected.

He once told me, "You've got the poise I never had." And maybe that's true. He didn't do well with bosses. He was entrepreneurial, always four steps ahead of people who didn't want to be questioned. The catcher behind the plate. That strategic mind never turned off. He'd ask me questions that cracked open new layers in my thinking. Things I hadn't seen, angles I hadn't considered.

If my dad showed me how to see the whole field, my mom showed me a different kind of strength.

Learning Quiet Strength from Stubborn Love

My mom, Patricia (Pat), was different. Quieter. Softer. The word people used was "sweet." But that was surface. There was steel in her. You just had to know where to look.

My mom,
approx. 8 years old

She came from a difficult childhood with years of abuse, instability, and being told to hide when anyone came to the door. When she was seven, her controlling stepfather decided he was "done with kids" and sent her and her younger brother to boarding school. No conversation. No goodbye. Just sent.

She didn't speak for the entire year. Not one word. The school labeled her as mute. Her silence wasn't weakness. It was protest. Even as a child, she refused to give something so personal (her voice) to people who hadn't earned it.

This taught me something fundamental about power and voice. Sometimes the strongest thing you can do is refuse to participate in a system that doesn't value you. Sometimes silence is the most powerful form of resistance.

Between the ages of eleven and fourteen, I went to my mom's work at an upholstery shop every day after school. I watched her work the high-powered sewing machine with breathtaking precision. Well, not always 100 percent precise. I watched her run the machine all the way through her thumb. Twice. She knew none of the guys would pull it out for her. Both times, she winced, muttered a rare "dammit," grabbed a pair of pliers, pulled the needle out, taped up her thumb, and went back to sewing.

This is where I learned ownership. Not just responsibility, but true ownership. When something breaks, you fix it yourself. You don't wait for someone else to solve your problems. You grab the pilers and take it out yourself. This mentality would later drive my insistence on internal capabilities over vendor dependencies. When systems broke at lululemon, we didn't just call the vendor. We learned to fix them ourselves.

She worked at that upholstery shop for eight years. She did every job in that shop except putting staples in and taking them out. She estimated jobs, rang every sale, answered every phones, served every customer, sewed every stitch, and wrote every check. She also knew that every man, including every day laborer who walked in off the streets, including the ones still drunk from the night before, was paid more than she was. Every guy, yes always a guy, who wanted to work the day pulling staples out of couches, made more than her. She managed payroll, so she knew. And she put up with it.

One day, she asked for a raise. Her boss gave her $.25 an hour increase. She finally asked him, "Why am I paid less than every other person who works here? I do everything here. Is it just because I am a woman?"

He replied, "Pat, that's just the way the world works."

She probably would have accepted that, but he continued, "*and Julie better get used to it too*."

And that was the day she resigned. She didn't make a scene. She didn't shout or write a letter. She walked away. When we got in the car I heard her fiercest swear word, reserved for rare moments just like this, "Bastard!"

She then opened her own upholstery business. She started from scratch, and she built something she was proud of.

This moment shaped my sense of fairness.

Her quiet defiance planted something in me. It became the voice I would later hear whenever I faced cultures that asked me to settle for less than I deserved.

Her high-end designer clients found her new shop, and her business flourished. She was able to pick the clients she wanted to work with and turned down those who weren't nice. She never hiked the prices. Working with good people was what she desired.

From her, I learned that integrity isn't just about honesty. It's about living a life that heals what hurt you. It's about creating the conditions for others that you wished you'd had yourself. Every time I've prioritized culture over convenience, every time I've chosen sustainable solutions over quick fixes, I've been building the kind of workplace Mom taught me was possible.

Building My Own Map

I was a nerd before being a nerd was cool. Throughout elementary school I sat in the front row of every class, shooting my hand into the air at every chance of answering a teacher's, devouring the learning. Being the first done with a test and getting a perfect score were my goals. Maybe it's clear already, but my social circle was very limited. As in, I had no friends. My mom said I came home from school every day in the fifth-grade crying.

It was that same year that my love affair with technology started when my dad enrolled me in a Radio Shack computer training class. I was ten years old, sitting in front of a TRS-80, learning BASIC programming, and suddenly the world made sense.

Here was a machine that did exactly what you told it to do. No hidden agendas, no social dynamics to navigate, just pure logic. If your code was right, it worked. If it wasn't, it didn't. The feedback was immediate and honest.

I was mesmerized. While other kids were outside playing, I was hunched over that computer, typing commands and watching them execute. My dad would find me there long after class ended, still experimenting, still learning.

By high school, couldn't wait to learn more. I challenged the 12th grade Computer Programming class as an 11th grader. I walked into that classroom and felt at home immediately. Yes, I was the only girl. I didn't care. I was there for the code, for the logic, for the pure satisfaction of building something that worked.

Outside of school I wasn't just playing. I was building real solutions.

Me in high school

At sixteen, I took a job as a receptionist at a construction company. The phones didn't ring much, which gave me time for homework. But what caught my attention was the computer. Early desktop computing. DOS operating system, command lines, floppy disks, Lotus 1-2-3. WordPerfect.

I started experimenting. Built automated spreadsheets to manage Accounts Payable and Receivable. Layered WordPerfect macros on top of Lotus 1-2-3 to generate employee paychecks, complete with prompts and tax calculations.

I was sixteen, maybe seventeen, getting paid to do something this fun. I started to get annoyed when the phone rang because it interrupted the work I really cared about.

I wanted more. At 17, I called a computer training company and positioned myself as an expert in desktop computing. They hired me to teach corporate classes. I created the curriculum and led hands-on sessions where I showed business professionals how to use productivity software.

Not long after, I was helping my dad with his consulting company. He'd install hardware and software for small businesses, and I'd come in behind him to train users or build custom solutions in Access or Lotus 1-2-3.

I loved the problem-solving. I loved creating systems that made people's lives easier. I loved learning about business.

The seed of solving business problems with technology was planted then. Not in strategy rooms or leadership seminars, but in small offices, hands on keyboard, building things that actually worked.

Technology became my refuge. A place where merit mattered more than anything else. Your program either ran or it didn't. While navigating systems that rarely made sense, I found freedom in one that was clean, clear, and honest.

So by the time I graduated high school, technology was already my path forward. But the journey to actually building a career wouldn't be straightforward.

I went straight to college after high school, but not the traditional route. I started at a community college, mostly because I was stubborn, and wanted to pay for my own college. My sister had spent years telling me, as the youngest, I had it easier than her and our brothers.

A baseball player didn't make much money in those days and they all had to pay for all of their own expenses, including college. I internalized that. Even when my parents offered to help me later in their lives, I turned them down. I needed to prove I could do it by myself.

But the independence I was building came from harder lessons too.

At sixteen, I was in a relationship that became abusive. What began with words and control grew into something darker. I stayed longer than I should have, partly because I didn't yet trust my own worth. That's what abuse does. It convinces you that you're the problem, that you deserve what's happening.

Walking away taught me that self-respect isn't negotiable. And it taught me to recognize a pattern I'd see again in corporate America too many times. Abusers test boundaries to see what they can get away with. No consequence means violations escalate. The faces would change, but the dynamic wouldn't.

My experiences taught me that silence protects the wrong person. Reporting isn't about causing trouble. It's about stopping patterns that will continue with the next person if you don't act. Predators rely on your hesitation, your fear that speaking up will make things worse. But waiting to see if situations improve on their own is how abuse becomes normalized.

These weren't just personal survival insights. They were leadership lessons I didn't yet know I'd need. The sixteen-year-old who learned to recognize control disguised as care became the executive who could spot toxic leadership before it destroyed teams. The seventeen-year-old who learned that power gets weaponized against the vulnerable became the CIO who acted swiftly when managers undermined psychological safety.

So I worked full-time and carried a part-time course load. I commuted, barely stopping to eat. Most days I went straight from work to class, studied late, got up early. There was no cushion. Just belief that moving forward was better than standing still.

After community college, I tried international business as a major. It didn't fit. I dropped out. A year later, I found an evening computer science program at Seattle Pacific University. I set a goal that I would complete my undergrad and earn an MBA by the time I was 30. During those years, I bounced between jobs that barely paid the bills.

Discipline matters more than motivation. Motivation fades, but discipline gets you to the finish line. Later, when leading eight-year transformations at lululemon, I learn to trust the process even when you can't see the outcome yet.

Self-reliance has a cost. I proved I could do it alone, but I also learned that refusing help isn't strength. It's often pride disguised as independence. Years later, when I built teams, I had to unlearn this learning. The strongest leaders aren't the ones who never need help. They're the ones who know when to ask for it and how to create environments where asking feels safe.

Looking back, those years of piecing together my education taught me that no one was going to hand me opportunities I hadn't earned. That building something slowly and deliberately beats waiting for perfect conditions. That stubbornness can be both asset and liability. It got me through, but it also isolated me.

Most importantly, I learned that the path doesn't have to look traditional to get you where you need to go. Sometimes the detours teach you more than the straight line ever could.

But knowing your path and walking it without compromise are two different things. The workplace would test everything I'd learned about standing firm.

When Disrespect Tries to Shrink You

In one of my early jobs, I had a boss who made it his mission to force me to choose between school and work. He didn't respect education. Called college something for people who "couldn't learn on their own," unlike him, the self-made genius.

One night, he invited me and another young woman from work to his house for dinner. Over the meal, he told stories about being a self-taught entrepreneur. The subtext wasn't subtle. The other woman had put her degree on hold. She had her priorities straight. He praised her for it. Not long after, he started scheduling me to work Saturdays. He knew it would force me to drop my classes.

He saw my commitment to finishing my degree as a threat. That was the first time I understood that disrespect doesn't just withhold recognition. It actively tries to make you smaller.

I called my dad, ready to quit in a spectacular blaze of fury. He gave me the long-game speech. "This isn't just about this moment. It's about all the moments still to come."

Least helpful thing I could hear right then. I wanted to call my boss out for trying to derail my education. Let him know how unfair and short-sighted he was being. Tell him I'd come back someday and prove him wrong.

But my dad, in his simple and level way, said, "Don't burn a bridge, even if it's on fire." Then, "Julie, give it 24 hours."

I woke up annoyed that he was right. Again.

I stayed. I finished my notice. I left on good terms.

On my last day, my boss walked me through the office like he was delivering parting wisdom. He stopped at my desk and said, "This will be the last time in your career you'll have an office with a door." Then, pausing by the breakroom, he added, "And the last time you'll ever have freshly roasted coffee beans. I just want you to enjoy this moment, Julie."

None of it came true.

I've had many offices with doors. Better coffee, too. But more than that, I built a career he could never have imagined for me. Not to prove him wrong. To prove to myself that I could keep choosing clarity over compliance. That I could trust my instincts and still move forward with grace. That I was already leading, even if no one called it that yet.

At another early job, I watched a female boss use manipulation as a power tool. She bragged about making older men uncomfortable in presentations, using their discomfort to get what she wanted. She got results. She also got a team that didn't trust her. They just managed around her.

You can't build respect by trading on shock value. When you rely on manipulation instead of skill, you're telling everyone, including yourself, that your intelligence and integrity aren't enough. You might win short-term. But you lose yourself in the process.

I was starting to see how it all worked. Disrespect as control. Performance as protection. The cost of hiding who you really are. These moments were shaping the leader I would become.

The Foundation Revealed

Looking back, my early years weren't about learning frameworks or methodologies. I was just learning to see things differently.

Twenty years after I left that job that my Dad coached me into leaving gracefully, I ran into my old boss at a café in Issaquah. He gave me a hug and told me how proud he was. He'd been following my career, seen the news articles.

Did I feel the vindication my 20-year-old self had dreamed about? Not even close. Apart from how things ended, he'd been a good boss. I'd learned a lot from him. He was a human doing his best to build a business and keep his people. All that steam I'd built up was gone. I was just glad to see him.

Turns out relationships matter more than being right.

That became the foundation for everything that came later. Not technical expertise, though that helped. Not strategic frameworks, though I'd use plenty. But the knowledge that systems shape people more than people shape systems. And everyone has potential if you care enough to look for it.

Before that foundation could hold anything, though, I had to learn what happens when you build without one. The protective mechanisms I'd developed—the corporate armor/masks, the performance, the careful editing—weren't just limiting my leadership. They were sabotaging it.

What to Carry Forward

Systems Thinking Is Your Foundation. When you walk into a complex organization, see the invisible dynamics. What incentives actually drive behavior? Where does information flow freely and where does it stick? Who has informal authority that doesn't match the org chart?

Quiet Strength Builds More Than Performance Ever Could. The strength that matters isn't the kind that announces itself. It's the kind that shows up consistently when no one's watching. Building capability while others manage perception. Doing the work that matters instead of the work that's visible.

The best leaders don't need the spotlight.

Discipline Beats Motivation. Motivation comes and goes. Discipline is what gets you through the years when the finish line feels impossibly far away.

Trust the process even when you can't see the outcome yet. The hardest transformations require staying committed long after the initial excitement fades.

Self-Reliance Has a Cost. Refusing help isn't strength. It's often pride disguised as independence.

The strongest leaders aren't the ones who never need help. They're the ones who know when to ask for it and how to create environments where asking feels safe.

The goal isn't to prove you don't need anyone. It's to build something bigger than what you could do alone.

The foundation my parents built gave me strategy and strength. But I still had to learn what happens when you try to lead while hiding who you are. That insight would take decades to learn, and it would cost me more than I knew.

CHAPTER 3

The Cost of Hiding

Your Personal Lens

This chapter isn't comfortable to write, and may not be comfortable to read. So why are we talking about trauma in a leadership book?

Because the capabilities that matter most in leadership—making sense of chaos, working without clear authority, staying grounded when everything's uncertain—aren't learned from case studies. You earn them by living through hard things.

And because the people you lead bring their whole selves to work, whether you acknowledge it or not. Trauma shapes how we respond to pressure, uncertainty, and change. The exact conditions you create when you're transforming an organization.

If you don't understand why someone shuts down during conflict, or works themselves into the ground, or can't take feedback without spiraling, you're missing half the story. And if you don't understand your own patterns—the ways your past shows up in how you lead— you'll keep hitting the same walls.

This isn't theoretical. You can't build psychological safety or lead through crisis if you're pretending everyone shows up as a blank slate. They don't. You don't either.

When I'm hiring senior leaders, I'd rather work with someone who has navigated real adversity than someone who has only known success. People who have faced hard things develop judgment, resilience, and empathy that smooth sailing never builds. They know how to read a room. They know when someone is struggling but won't say it. They know the difference between a setback and a crisis.

That perspective becomes their advantage.

Years later, I would fire toxic leaders who were delivering strong business results. I would push back against systems that demanded I stay quiet about what wasn't working. I would build cultures where people felt safe to surface problems before they became crises.

All of it started with that moment when I learned my first lesson in trauma at 16.

Learning What I Wouldn't Accept

I was 15-1/2. A 17-year-old guy, outgoing and charming, started paying attention to me. I had just transferred to the state's second largest high school from a very small private school. I had no friends and felt like a very small fish in a very large pond. I was grateful for someone talking to me. We shared a debate and competitive speaking class, so we met outside of class to prepare for weekend tournaments. At last I had a friend.

When I wasn't interested in dating him, he decided I must be "closed-minded," and spread the rumor around Auburn High School that I was "a prude from a private school." If I wouldn't date him, there was clearly something wrong with me. Oh how I wish I could go back to the 15-1/2 year old version of myself and hug her and tell her that she was enough.

The rumors isolated me even more. He kept asking. Eventually I agreed to go on a date with him.

What followed wasn't a relationship. It was control. Then violence.

I wore long sleeves to hide bruises. I stayed silent because I'd seen what he was capable of. When an exchange program to Brazil offered escape, I jumped at it.

As we walked toward my plane at SeaTac, my parents were on my left side and he was on my right. He held my hand and squeezed it so hard it swelled. I iced it on the plane. His last message of control, but my first step toward freedom.

Until the phone rang in my host family's living room in São Paulo.

He'd found me. Called again and again until my host mother locked the phone and began to doubt me. She thought maybe I was returning his calls, racking up international charges. I wasn't. I begged him to stop calling to no effect. Then, one night, the phone rang as I was sleeping. My host mother came to wake me up. He didn't even start with pleasantries. He said simply, "come home, or I'll burn your parents' house down."

I'd seen him smash mailboxes, slash teachers' tires. I knew the darkness inside him. I believed him.

I came home. But I didn't come back the same person. I came back with perspective on who I truly was and a commitment to get back to that person. I was returning to the person had strong integrity, was stubborn as hell, and wanted to create her own future.

Walking away from that relationship taught me that self-respect isn't negotiable. That insight would serve me well in corporate America, where I'd face different kinds of control, different expectations that I shrink myself to fit.

Turns out he'd made all those calls from local churches, asking each one to borrow their phone for a quick call. The phone company pointed the bill back to me. So I got a job at Wendy's to pay it off.

One night, the manager followed me into the walk-in freezer and pushed his body up against mine, my back pressed against the metal rack. I shoved his shoulders with everything I had, twisted out of his reach, and walked out.

The next day, he gave me an absurd punishment, which was to scrub out the *inside* of the grease bin behind the store.

I picked up the bucket and brush, filled it with soapy water, walked straight out the back, set everything down beside the grease bin, got in my car, and drove home.

Then I reported him.

He was dismissed. I went back to work. Every day I was scared he would come back. I checked the parking lot for his yellow convertible Corvette before each shift.

Now I understood viscerally what that third-grade neighbor girl had felt when she called me, begging me to stay quiet. The fear wasn't about the abuser, it was about what happens when the system gets disrupted. When you've learned to survive by managing someone else's violence, the person offering help feels more dangerous than the person causing harm.

But I also learned that the only way out is by changing the situation. Safety doesn't come from managing the abuse better. It comes from refusing to participate in the system that enables it.

My story is one of many. Anyone who has less power in a system knows these dynamics. And every time someone speaks up, the world moves a little. Change begins when one person refuses to stay quiet.

A Decade of Hiding

By the time I got to college, I'd already learned to split myself. To show the parts that felt acceptable and hide the rest.

Studying computer science, in my first real algorithms course, I was amped up. I wasn't just playing. I was building something real. My brain was alive.

But slowly, something shifted. I was the only female in most classes, and it started to matter. Comments from classmates weren't harsh, but they made it clear I didn't quite belong. As I walked to the class, a guy in the back asked, "Are you sure you're in the right class?" Maybe he meant it as a joke, but I felt myself pull inward.

I tried to make friends, but conversations drifted to dating. Small talk that didn't feel small anymore. Beneath it was a realization I hadn't named yet, to anybody but Cindy. I didn't want to date the boys in the room. I didn't want to date boys at all.

So I tried to disappear. Show up right as class started. Leave as soon as it ended. I only remember one name. He asked me out, and when I said no, he was kind about it. A glimmer of ease.

I didn't know how to name what I was feeling, so I didn't name it. I just got quieter.

That's how retreat works. Not with a blowout moment, but with small choices to stay quiet.

At nineteen, when I started dating Cindy, I was excited and happy. She was the calm, kind and consistent to my restless energy, my sharp edges, my urgency about everything. But I didn't know how to share that joy. I believed we needed to stay hidden if I wanted a career. So I hid. Because when you don't feel safe being fully seen, you learn to split yourself.

For over a decade, from my late teens to my early thirties, I lived inside a carefully built box. The walls were fear, the ceiling was shame, and I kept reinforcing both.

At Weyerhaeuser, I had an assistant named Sharon. Her actual title was "secretary" back then, which included Friday martini lunches. She was warm and curious and genuinely wanted to know me. She answered my phone when I wasn't there. Imagine that. Phones tied to desks.

Sharon cared. And I was afraid to let her.

I saw a domino chain. Sharon learns about Cindy, tells another assistant, word spreads, and suddenly I'm exposed. So I stayed at my desk. I asked Cindy to only call when I wasn't in meetings. I made the person I loved adjust to my fear.

When I left Weyerhaeuser in 1998 for a new role at fine.com, I thought everything would change. This was the early Internet. Hip, modern, downtown Seattle. Exposed ceilings. Live music at company parties. The kind of place where being different felt like part of the brand.

But I still kept the shell. Mondays always started with "What did you do this weekend?" I learned to speak in careful riddles. Never "we." Always "friends and I." I edited Cindy out of every story. It made my life smaller than it was.

The cost wasn't just professional distance. It was personal erasure.

One evening, sitting in my parents' living room, my brother said he'd canceled the newspaper because they printed a pro-gay-marriage op-ed. Cindy and I sat right there, invisible, while my family debated whether people like us deserved rights.

I didn't push back. I let the moment slide by. I told myself silence was safer. We felt small and immaterial, but stayed because the security of having family was more important than being seen.

Looking back, I can see what I was really doing. I wasn't only protecting myself. I was protecting everyone else from having to rethink what they believed. They got to stay comfortable. Cindy and I paid the price.

And in the process, I lost out on connection. On allies. On people who would have understood. I was so focused on avoiding discomfort that I forgot to value my own humanity.

The hardest part? Without meaning to, I was modeling that authenticity is risky. Keep the real parts tucked away. Belonging is earned through editing.

I was helping create the very culture I would later work to dismantle. A culture where psychological safety is for after hours. A culture where truth stays quiet.

The box I built to survive also held me back. But it didn't stay intact forever. Cracks started forming. Light started getting in. And even though I couldn't see the change yet, those cracks were the beginning of freedom.

Breaking Free

Our oldest, MacKenzie, was speaking in full sentences when my mother brought it up in her kitchen. We were celebrating my brother's anniversary, and she turned to me casually and asked, "Do you and Cindy have a date… that you celebrate?"

My first instinct was to look for a fire alarm to pull. Run. Deflect. Change the subject. I'd spent over a decade hiding, and I didn't know what would happen when that wall came down.

But instead, I stayed in the conversation.

Within minutes, we were both crying. Not because it was painful, but because she was relieved. She'd been waiting all those years for me to say something. She'd known and had been giving me space to tell her when I was ready. She later told me, "I knew about Cindy from the first time you described her on the phone to me."

This was the same mother who, in the second grade, stopped speaking for an entire year when she felt unsafe. The same mother who taught me that silence could be protest, or protection, or survival.

I was her daughter. Of course I'd learned to go silent too.

But on the other side of that conversation, I felt something I hadn't felt in years: *safety*. Not just acceptance but belonging. I still felt exposed at work, with colleagues, with the professional identity I'd built so carefully. But the possibility of losing my parents was gone. They saw me, and I was still their daughter.

I walked out of that day feeling like the world was a different place. One that I belonged in.

The very next day, again in my mom's kitchen, MacKenzie proclaimed to one of my mom's friends, "I have two mommies!" I knew at that moment that I couldn't set an example for her by hiding inside our family. She deserved to be her full self, and that required the same of me.

That conversation didn't fix everything. I still hid at work for years after that. But I had just experienced the fact that being seen didn't destroy the relationships that mattered most. That the fear I'd been carrying about rejection was worse than the reality of truth-telling.

Years later, when Mike Richardson, my mentor, would ask me how my team could follow me if they didn't know who I was, I'd think back to that day with my mother. To the relief on both sides of honesty. To the way silence had been protecting me from a loss that was never going to happen.

The armor I wore at work wasn't just about career protection anymore. It was about a habit I'd perfected but no longer needed.

When Everything Looked Solid

After the conversation with my mom, everything looked solid. I was even began talking with my co-workers about Cindy.

I was a vice president at a thriving ad agency in Seattle. I had built a high-performing team, hand-picked and deeply trusted. We were doing award-winning work. I loved the culture. I loved the people. I loved the pace. It felt like a good time to expand our family, so Cindy and I made the decision for me to have a baby.

My boss, John, was all charisma. A big-voiced creative who turned every hallway into a stage. On February 28, 2001, the day of the Nisqually earthquake, John and I were talking in his office when the whole building began to sway. Without missing a beat, John threw his arms wide and shouted, "Come on, Smith Tower. Come on Baby, hang in there!"

When the shaking finally stopped, we discovered we could slide pencils through the new gaps between our offices on the 26th floor. Later we laughed about how differently we'd each handled the same crisis with John cheering on a 90-year-old skyscraper and me trying to enforce earthquake protocol while the ceiling literally split above us.

Together we were a great pair.

Then the agency was acquired. We were in the middle of the Dot-com crash, and our new parent company was looking for ways to build efficiencies between offices.

The new executives flew into Seattle to meet the team. Over lunch, they bonded over their German cars—they even had a name for it, the GAC (German Automobile Club). They were all in blue suits, talking cars and club memberships. I was pregnant, in stretchy pants, and I could already feel the gap.

They asked me to fly to Boston. The night before the meeting, they'd chosen a cigar bar. The smell made me cough. I nursed a sparkling water, mostly just wanting to go back to my room and elevate my swollen feet. I was carrying a growing human, exhausted, and completely out of place.

The next day, we gathered around a long conference table. It felt like they all knew each other. I was the outsider. The only other woman in the room was the head of HR. Twice I tried to contribute and both times I was talked over. So I stopped trying. I sat back, stayed quiet, and counted the minutes.

A month later, I was laid off. Not exactly surprising. But still crushing.

From Being Wronged to Leading Forward

Six months pregnant. Laid off. The breadwinner in our family, suddenly with no income and a baby coming in three months.

I told the story to everyone who would listen. Cindy heard it over and over. My parents heard it. My friends heard it. Each time I told it, the narrative got sharper, my victimhood clearer, my righteous anger more justified.

I hired a lawyer. Not because I thought I'd win, but because I needed to be heard. And when I saw John's emails during discovery, I felt destroyed. He had called me "expendable."

Because in my quest to be right about being wronged, I'd lost something more valuable than the job or the severance. I'd lost John. Not the boss who'd laid me off, but the person I'd huddled with in a doorway during an earthquake, laughing while the building swayed around us.

Years later, in a leadership development session, the coach asked us to think of a time when we'd been truly wronged.

I told my pregnancy layoff story. Six months pregnant, laid off, blindsided. The cigar bar where I couldn't breathe. The conference table where I was talked over. I had told the story so many times it became a script.

But then she asked us to do something different. To shift to an empowered mindset and tell the story from the angle of what we owned in that, and what we could have done differently.

It hit me like a fast-moving train.

I saw the missed openings. I could have asked about their cars, not because I cared about horsepower, but because they were handing me a point of connection. I could have skipped the cigar bar, said "I'm pregnant, I need rest" to show up fresh the next morning, curious and ready to contribute. I could have asked John what support he needed and, if the layoff was still inevitable, sought to understand how to position myself for what came next.

I had let fear run the show. I'd lost my sense of myself, my confidence.

But it also marked a turning point. It broke the illusion that achievement alone would keep me safe. And three months into it, I was able to fully experience the joy of a newborn. I didn't have the pressures of working. I found I loved being a mom. Our daughter MacKenzie was the best gift we could have been given, and I had the opportunity to spend the first year of her life with her.

It was a privilege of having savings built up and a wife who could work during that time. I look on it now as a time I will never regret.

It was also the beginning of a different kind of strength…the kind that quieter, steadier, and rooted in perspective. I was learning that pain can be a teacher if you don't let it become your identity.

Hard Feedback

The first exit interview landed in my inbox on a Tuesday. My HR partner forwarded it with a note that simply said, "We should talk about this."

I read it as my entire office suddenly felt chilly. The words were clinical. "Julie isn't present, doesn't really know what's going on," "seemed disconnected from the team," "unclear if she cared about outcomes or people." But the message was clear. I was failing as a leader, and everyone could see it except me.

My first thought was that again, people didn't understand me. I certainly care. How do they not know this? I was annoyed at him for not talking to me directly and wrote the incident off.

But then, two months later, a second manager resigned. Different person, same feedback. "Julie is smart, but I never felt like she knew me. I never felt like I could bring problems to her without having them minimized. She didn't have my back."

I sat down, closed my office door, and stared at the words until they blurred.

Two managers. Two exits. Same story. Two people who didn't understand me. Or, maybe they did. What if they were right? I began to see the wall I'd built so carefully. But that was protecting me. Without it, I'd be _vulnerable_, a word that terrified me.

I had spent so much energy hiding who I was. I was hiding Cindy, hiding uncertainty, hiding any vulnerability that might be used against me. The wall I thought was protecting me was preventing me from being a good leader. I'd forgotten how to *actually connect* with people. The corporate armor—the masks—I thought were keeping me safe were making me impossible to work for. The good news was simple but not easy. If I had built the wall, I could be the one to take it down.

That night, I called my Dad. It wasn't unusual. I called him every night. He heard my heartbreak and then said, "Now you know. What are you going to do about it?"

When Competence Becomes a Cage

The masks I thought were protecting my career were actually constraining it. Turns out, when you edit yourself to make others comfortable, you teach everyone that editing is the right thing to do, and authenticity is dangerous. And more than that, you're just performing competence. You're creating environments where problems hide until they become crises.

In leadership meetings, I noticed people over-preparing instead of pressure-testing ideas in real time. Things were rehearsed. Everything was safe. No bold proposals that might fail. They said what they thought I wanted to hear instead of what they believed.

I wasn't just keeping people from knowing me. I was keeping them from bringing their full capabilities to the work.

The feedback from those two managers shattered my illusion that hiding was just about me. My choices were shaping the culture, whether I intended them to or not.

I had to stop performing competence and start showing up as human. That meant admitting when I didn't know something. Mentioning Cindy naturally instead of editing her out of every story. Asking for help instead of pretending I had it all figured out.

The response surprised me. People didn't see weakness. They saw permission to be human, to struggle, to learn out loud. When people feel safe being human, they bring their full intelligence to work.

The business impact was immediate. Problems surfaced earlier. Collaboration improved because trust replaced performance. Innovation accelerated because psychological safety enabled risk-taking.

Most importantly, I started attracting different talent. People who wanted meaningful work in environments where they could be themselves. People willing to challenge assumptions because they trusted their voices would be heard.

The authenticity I'd feared would cost me everything became the foundation for trusted global teams, technology transformations, and cultures of inclusion that became competitive advantages.

Using the Platform

Years later, while at lululemon, we gathered for what would be our last in-person Leadership Summit, only we didn't know it at the time. Every February, global leaders and store managers from every market, came together in Vancouver to align on strategy and strengthen our culture. Except this was February 2020. At the last minute, the China team sent regrets. COVID had hit them first. We proceeded anyway, not yet understanding what was coming.

We wrote postcards to China telling them we missed them and hoped we could see them soon. We had no idea.

In 2021, at what would be our first virtual Leadership Summit, we want to make this days-long event meaningful. Instead of hosting a headline name to be a guest speaker, they asked me. I thought about it. This was the ultimate exhibition of vulnerability I could have imagined. But a friend said to me, "You have the platform. Use it."

So I did.

I told my story—the hiding, the fear, and the long road to being fully myself. I spoke about finding refuge in technology, about being the only woman in the room, and about the mentor who finally asked, "How are you going to get people to follow you if no one knows the real you?"

The polished video didn't feel real. I deleted it and did one honest take. (Full speech is included in Appendix A.)

When it aired, I braced for regret. Fear. Rejection. Instead, my inbox filled with messages from people all over the world—people who had been hiding parts of themselves, too. That's when I understood that telling my story wasn't risky. It was leadership.

Leading at this scale had taught me that authenticity doesn't shrink at scale. It amplifies. And trust scales faster than any system we'd ever built.

After a year of leading through crisis from kitchen tables and home offices, after watching people show up with toddlers in the background and pets interrupting standups, after navigating grief and fear and exhaustion together—the walls were already down. My story just gave everyone permission to acknowledge what we'd all been living. We now knew that bringing our whole selves to work wasn't a liability. It was what had carried us through.

What Trauma Taught Me

Looking back now, I can see how those experiences, as painful as they were, created an unconventional leadership education.

I learned to spot micro-signals others missed. When someone's voice got tight in meetings, when body language didn't match words, when enthusiasm felt performed. Years later, I could identify team dysfunction before it exploded.

Having power stripped away taught me how to wield it carefully. I learned to ask questions instead of issuing orders, to distribute decision-making instead of hoarding it. People knew they could bring me problems without becoming the problem themselves. Fear shuts down the truth-telling you need for transformation.

I understood what it actually takes to keep moving when everything is falling apart. This meant I could stay calm during system outages, maintain perspective during crises, and support team members through their own impossible moments. I knew what "impossible" looked like, and most business challenges didn't qualify.

I developed the ability to sense when others were struggling before they asked for help. I would check in during brutal project timelines, recognize when people needed backup, and eliminate toxic managers quickly. This wasn't about being nice. It was about protecting my investment in talent and removing threats to performance.

Boundaries became operational tools, not theoretical concepts. I could recognize toxic situations quickly and act decisively instead of hoping they'd improve. This would save my teams from extended periods of dysfunction that other organizations endured because their leaders couldn't make hard calls.

These were capabilities forged under pressure and refined through practice. And they would prove more valuable than any advanced degree when everything started breaking at once.

For years, I thought strength meant having all the answers. Never showing doubt. Never admitting mistakes. Never letting people see you struggle.

I was wrong.

The strongest leaders I know are the ones who can say "I don't know" without losing credibility. Who can admit when they've changed their mind. Who can show up as fully human without compromising their authority.

The most powerful thing I ever did wasn't coming out. It was coming home to myself. Because when you stop editing your truth to make others comfortable, you stop disappearing. You start leading from alignment instead of fear.

And that's when real transformation becomes possible.

This isn't only a personal truth. Organizations mirror the same pattern. McKinsey's global research on diversity and inclusion found that companies in the top quartile for ethnic and cultural diversity were 33-36 percent more likely to financially outperform their peers[2]. Inclusion is not just about fairness; it is a driver of performance. Cultures that welcome more voices consistently unlock better results.

The masks weren't protecting my career. They were limiting it. And the moment I stopped hiding, I started leading.

For Your Leadership

Years of hiding taught me that silence protects the wrong person and masks limit your impact.

Truth-Telling Is Your Superpower: When you see a pilot failing or a deployment creating bias, your voice can redirect resources before they're wasted. The pattern of truth-telling becomes the culture that catches problems early.

Authenticity Unlocks Intelligence: When I stopped hiding, everything changed. We recruited better talent, moved faster with innovation, built stronger culture. AI transformation needs this even more. People bring their full intelligence and energy only when you show up as fully human first.

Connection Comes Before Credibility: In transformation, people won't follow based on your technical expertise. They'll follow based on trust in your judgment, values, and commitment to them. Know your people. Let them know you.

Vulnerability Builds the Foundation: When I stopped performing competence and started admitting uncertainty, problems surfaced earlier and collaboration improved. If your team is afraid to say "I don't understand how this model works," you're building AI on a foundation of fear. When you can say "I don't know" without losing credibility, everyone learns out loud. That's where breakthroughs happen.

The patterns that saved me personally, including speaking truth, building trust, and refusing to hide, became the patterns I used professionally. But first I had to learn them in organizations that gave me a front-row seat to what happens when technology and business work together.

NORDSTROM
NORDSTROM.COM
STORE
DIGITAL
COMMERCE
WEBSITE

CHAPTER 4

Nordstrom: The Business of Technology

The lessons I would later use as transformation patterns weren't invented in boardrooms or learned from business books. They were forged in crisis, tested under pressure, and refined through years of leading change in organizations that weren't ready for it.

Nordstrom: A Different Kind of Company

Bruce Nordstrom, father of Blake, Pete and Erik, and uncle of Jamie, joined my evening University of Washington MBA class in 2000, spending three hours talking to our small class of 29 people. I was so impressed that this accomplished man would take his entire evening and spend it with us. The stories he told, from his family building a customer-obsessed business by kneeling in front of the customer and selling one pair of shoes at a time, to his decision to take the company public, to returning tires for a customer because a Nordstrom store was built where the tire store used to be opened my mind to a different kind of leadership. A humble, quiet, people-centered leader. He was different and he was genuine. I was struck.

When I had the opportunity three years later to join Nordstrom, I jumped at the opportunity. And I landed in one of the most incredible transformations of the time, a 100-year-old retailer integrating its catalog and e-commerce business into what would become one of the first dominant omni-channel retailers.

At that time, I had just taken a job as an adjunct professor at Seattle University. I was planning to get my PhD and teach at a university. After getting laid off while pregnant I was disillusioned about working in a corporate environment ever again. However, when Nordstrom called, I decided to take the contract job and shifted teaching to the evenings. Three months, build some cash reserves, then ramp up teaching, that was the plan. I held both jobs for four years and stayed at Nordstrom for a decade.

I soon learned this corporation was different from anything I had experienced, or even imagined, before. People mattered. Tenure was expected. People stayed because they loved the company. The culture was palpable and resonated with me to my core. I was motivated to do incredible things for this company.

The Cultural Integration Challenge

I joined Nordstrom Direct, which at the time was a separate company from Nordstrom, Inc., and its brick-and-mortar stores. Nordstrom Direct was the catalog business with a newly-developed website, nordstrom.com. My challenge was being part of a team chartered with integrating these two companies into one and understanding the opportunities that provided.

The first challenge was culture. Nordstrom was a fashion business where typical attire was suits and dresses with heels. Nordstrom Direct was the opposite. Birkenstocks and shorts were the fashion. One of the biggest points of early contention was the mandate from corporate to wear close-toed shoes. That was soon followed by the no-jeans, no shorts mandate. I thought we were going to have a mass exodus of talent then and there. We did lose some folks, but most adapted. Fundamentally, the culture of service was common. And the technical challenge was enticing.

We built the first mobile app for Nordstrom. It was pioneering. Things we know now as common knowledge, we were learning the first time. As we turned on new features to the website, we learned the impact of pixel placement on conversion. No A/B testing, this was early days and there was no roadmap. We got to brainstorm, to build, to learn, to carve new paths.

But while I was learning to see systems in the technology and business, I was still blind to the system of influence.

When I was invited to join the CIO core team, I thought it was recognition of my technical capabilities. I didn't realize it was also a test of whether I could show up as a leader among peers.

Learning Influence

After one too many times watching myself vanish, I decided to try a different approach. If I couldn't yet figure out how to show up authentically, maybe I could at least figure out how the system worked.

I'd been invited to join the CIO core team. Six of us, all senior technology leaders, sitting together in a glass-walled office they called the Shark Tank. The idea was transparency. Leadership in full view.

I was the only woman. My desk looked straight into the men's bathroom.

Two truths about that experience. One, I never quite fit in. And two, I never let myself try.

One-on-one, I worked well with each of them. But as a group, they had rhythms I was still learning. They challenged each other easily, navigated the unspoken dynamics of who had insight and who didn't. I watched it all, trying to understand how influence truly worked.

The Enterprise Architects came in one day to present their vision for where Nordstrom was headed. Bold pronouncements about the future of retail technology. I thought they were missing something critical.

I had a framework in my head that could have shifted the conversation. Questions that would have sharpened their assumptions. A perspective that might have helped.

I sat there. Silent.

I didn't yet understand that being right doesn't matter if no one hears you. Having the best insights means nothing if you can't influence the room.

That pattern played out more than once. I'd prepare thoroughly. I'd have insights that could move things forward. Then I'd wait for the perfect moment that never came.

I thought intelligence and preparation were enough. They weren't.

Then I had a project that needed funding. A significant infrastructure investment that would enable the omnichannel experience our customers were starting to expect. The business case was strong. The ROI clear. The need, urgent.

I prepared a presentation that laid it all out. Data. Projections. Risk analysis. I walked into the executive funding committee confident that the strength of my analysis would carry the day.

They denied it.

I was surprised. The numbers were solid. The strategy sound. My thinking was right. But being right hadn't mattered.

This time, I had to understand.

After the meeting, I did something I hadn't done before. I asked for help. I went to a senior leader I trusted and said, "Help me understand what just happened."

He gave me one of the most valuable insights of my career. "Julie, decisions like this don't get made in the funding committee. They get made before anyone walks into that room."

He explained what would change everything for me. The formal meeting wasn't where influence happened. It was where influence got confirmed. If I wanted funding approved for something that was going to change the company, I needed to build conviction first. One conversation at a time. Understanding each stakeholder's concerns. Adjusting the approach so they could see them-selves in it. By the time we walked into that room, the decision should already be made.

The best analysis in the world loses to mediocre anal-ysis backed by strong relationships. The smartest person in the room has less impact than the person who knows how to bring others along.

I'd been playing the wrong game entirely. I thought if I just worked harder, prepared better, had sharper insights, eventually people would listen. But organizations don't work that way. Influence isn't about having the right answers. It's about helping other people arrive at those answers with you.

So I learned to work within the system. Pre-meetings before the meeting. One-on-ones with each stakeholder to understand their concerns and let them shape the approach. Building a coalition before I ever asked for a formal decision. It felt time consuming and inefficient. But from that day forward, 100 percent of my funding requests were approved.

Not because my analysis had improved. Because my influence had.

I was learning. I could navigate the organization now. I could build support. I could get things approved. These weren't small skills. They were essential to making change happen at scale.

I'd spent years thinking leadership was about being the smartest person in the room. I was learning it was about making the room smarter.

But I was also starting to see the next layer. I'd learned the mechanics of influence, but something deeper was still missing. I could work the system brilliantly behind the scenes, but I still wasn't showing up fully in the moments that mattered. That lesson was coming, just not yet.

Learning from Crisis: The Anniversary Sale

I learned the importance of Peak events through pain. Nordstrom has an annual event called the Anniversary Sale that its most loyal customers time their vacations around. The retailer marked its newest products down for a short period before marking them back up. Nordstrom wasn't a markdown company, and other retailers only marked down products that weren't selling, so this was special. It was one of the company's highest volume periods of the year.

Catalog customers were shifting to nordstrom.com instead of the call center. On the first day of Anniversary Sale that year, the website crashed. Hard. Too many customers hitting at once. We were running on an AS-400 system called Ecometry. It was on the roadmap to replace, but that didn't matter when customers were frantic and it was offline.

The website crash sent loyal customers to the phones. They knew inventory would sell out quickly. Years of filling out paper forms and mailing them in had taught them to act fast or the best items would be gone. Phone lines were clogged, up to three hours to reach an agent. And they waited.

I didn't know at the time that this emergency would repeat itself a few times in the future, each one with a clear lesson.

This crisis, though, was the first teacher. The Call Center put out an SOS for more hands. I took everyone on my team who wasn't fixing the website to the Contact Center, 20 minutes north of Seattle. We hopped on the phone lines. What surprised me most was how grateful customers were that we answered.

My first call started with, "Thank you for calling Nordstrom, this is Julie." I was hoping I could figure my way through this system I supported but didn't operate. How hard could this fundamental system be? Oh wait, F9 for this, F6 for what?

"Oh, thank you for picking up. How are you doing?" in her Southern drawl, she was concerned about me and could tell I was nervous.

"I'm so sorry the website isn't working. Thank you for your patience."

"Don't you worry. Every year I buy my grandkids the latest pajamas from the Nordstrom catalog, and I was so scared I couldn't get them this year, but now you're here." I felt it. I wanted her to get those pajamas. She told me about her grandkids and how Nordstrom fit into her traditions.

Years earlier, Bruce Nordstrom had told our MBA class about building a customer-obsessed business by kneeling in front of customers and selling one pair of shoes at a time. I'd thought that was a metaphor. Turns out I'd be doing it myself, literally—sitting in a call center during our biggest crisis, helping a grandmother get pajamas.

I learned the importance of being in the moment with the customer. It also hit me that she wasn't the same customer walking into our stores. She wasn't interested in top fashion trends, but comfort. She ordered her annual fish print loose-fitting dress, modernized each year in a new color. PJs and the fish dress. She was loyal.

At the time, merchandise sold online shared little overlap with stores. Catalog and nordstrom.com items were loose-fitting with a lower chance of return. They were classic styles repeated year after year, rather than fashion that changed seasonally.

This crisis taught me three things that would shape every transformation I'd later lead. First, technology failures have human consequences. Second, you lead by showing up where the problem is happening. And, third, the people closest to the crisis often have the clearest view of the solution.

The lesson was bigger than a sale. It was a view into the roadmap ahead. Combining inventory systems meant disappointing some customers as we walked away from styles they were loyal to. This wasn't a technology journey anymore. This was fundamental business transformation. And the crisis showed the urgency of getting it right.

Crises reveal the shape of your system. When our website crashed during Anniversary Sale, we learned that our real infrastructure wasn't the servers or the code. It was the people who answered phones, the trust customers had in our brand, and the willingness of everyone to jump in when things broke.

The tech keeps changing—from AS-400s to AI—but the lesson doesn't. It holds that if you don't build systems with deep respect for the people they impact, you're just building a faster way to fail.

The Omnichannel Opportunity

The biggest technical lesson was that our e-commerce engine would not perform under load. *Literally, too much load.* Ecometry was housed in a downtown Seattle building with floor weight capacity constraints. The system was too heavy to add onto! The fire was lit to replace Ecometry.

We built a plan to merge Nordstrom Direct's systems into the main company. Four phases, each one riskier than the last. Phase four, the final phase, would integrate the full inventory and commerce systems, replacing Ecometry. But it was near the end of this project when serendipity met preparation.

And it was the business opportunity hidden within a technical constraint that showcased systems thinking—and taught me that technology transformation requires business transformation in equal measure.

An Unexpected Business Breakthrough

All online orders were shipped to customers from a warehouse in Cedar Rapids, Iowa. *Except cosmetics.* Beauty goods weren't conducive to warehouse fulfillment—expiry dates, limited supply, recall requirements, fragile items, hazardous materials shipping, strict vendor contracts—so they were filled manually through a single store. The manual process was not scalable for the company's growth.

So, the final phase of the omnichannel transformation needed to not just replace Ecometry, but automate cosmetic orders across all stores.

Then the reality of what this meant hit like cold water in the face. One of my leaders and I were discussing cosmetics when she said, "We have to build the ability for stores to ship products strictly because of cosmetics, and it's a lot of work." I clarified, "But are we doing more work to filter cosmetics than it would be to just sell all inventory?" She confirmed yes. We both realized we were missing a huge opportunity.

I called an emergency meeting with Jamie Nordstrom who was President of Nordstrom Direct at the time and the COO. I explained that technically we could make all inventory available online and increase sales, but we had no time to understand impacts to stores or do proper training. We needed to decide quickly so we could sunset Ecometry before the next Peak season, which was approaching fast. We knew Ecometry was near end-of-life. We were about to find out how near.

"Should we continue building the filter for cosmetics or should we sell all company inventory?" I asked.

Jamie looked at me as if I had two heads. "Is this a question?" I reiterated that we didn't have time to understand the impact or plan for change at the stores. The

COO interrupted, "The stores will react. Let's do it." And with that, the decision was made.

The day *before* we were set to go live, Ecometry had a total hardware failure. Lights off, no heartbeat. We later joked it knew we were coming for it and wanted to go out on its terms, but in the moment we were panicked. We urgently began the transition to the new system. Luckily, we'd already converted all data and were just waiting for the green light. Through fate, it became September 13, 2009, and we thanked our lucky stars we were ready. The launch of the new system went off without a hitch. At least on the technology side.

The Day Everything Changed

On September 14, 2009, I went to the downtown Seattle store to watch the first order come into the store. It would go to the customer service office and they would print individual orders on the register tape. "MAC FNDTS 03TAN" meant a beauty item from the vendor Mac in a color of 03Tan. It worked for cosmetics. We were ready.

Then the orders started pouring in.

Jamie called me from the Chicago store, "Julie, we have a problem. I think Chicago is getting the entire company's orders. I'm sitting here with 21 feet of register tape filled with individual orders." I was seeing a similar thing in Seattle, but the orders were unique to each store. He said, "What do I do?"

I suggested he start filling orders.

I did the same. My first order was for a pair of socks. I had no idea where to start looking in the 3-story, 383,000 square-foot building. The order on the register tape didn't tell me which department it was in. The three-digit color code of AST didn't even make sense to someone who couldn't translate that to amethyst. And I wouldn't have been sure what that color was anyway. I realized my time was better spent with the team figuring out how to solve the problem we had just created.

Every store was overwhelmed with orders. Every person became part of the fulfillment network.

What We Hadn't Solved: The Business Transformation

The technology worked beautifully. The business operations were in chaos. That was the risk we'd taken together, but it didn't ease the front-line mayhem.

Within hours, store managers started calling. "My associates are abandoning the sales floor to pack boxes. Customer service is slipping." Another, "my best salesperson spent four hours fulfilling online orders today. Her numbers look terrible, and she's worried about making her target."

Then the customer calls rolled in. "Why did my order cancel?" "Why did I get three packages for one order?" "My shirt arrived with a security tag still on it." Fulfillment dropped to 50 percent. One in two items reached the customer.

This hundred-year-old business pivoted fast. Store Operations created new processes to pick and pack efficiently. Mailrooms added checks before shipping. We put in simple inventory filters that cut low-value or low-stock items. Fewer orders came through, but the ones we shipped were right. Fulfillment climbed back into the high 90s.

Meanwhile, we had changed what we were asking store teams to do. Associates had always been measured on personal sales and customer service. Now they were spending hours fulfilling online orders, work that didn't count toward their compensation or career goals. The incentives didn't match the strategy, and behaviors followed. Some associates hid high-end merchandise under the counter to reserve it for commission-earning in-person sales.

Over the next months, we partnered with Store Operations, HR, and Finance to retrofit the business transformation we should have designed upfront. We updated metrics so fulfillment counted. We aligned compensation with the new work. Managers learned that serving customers "in the back" was still serving customers. Regional leaders finally saw productivity across both in-store and online channels.

And it worked because each function owned their piece. Store Operations owned the workflow changes. HR owned the compensation changes. Finance owned the reporting changes. Technology enabled all of it with data and scale, but we couldn't solve it alone.

I learned my job wasn't only to build technology. It was to surface these dependencies early and make sure we were solving the whole problem, technology and operations together, from day one.

The Partnership We Should Have Built Earlier

Looking back, the mistake wasn't the speed. It was assuming that technology readiness meant organizational readiness. We knew the risk, but we didn't really understand its consequences.

We should have involved Store Operations from the moment we realized all inventory could go online. Not to review our technical approach, but to co-design the solution. They would have immediately seen the compensation misalignment, the training gaps, the workflow disruption.

We should have brought HR in early to redesign how success was measured. Not as an afterthought when behaviors went sideways, but as a core part of the transformation design.

We should have partnered with Finance to model the revenue attribution before launch, not scramble to build dashboards after the fact.

It was here that I really saw the pattern that would become foundational to every future transformation. Technology transformation is never just a technology problem. It's an organizational transformation problem that requires partnership between technology and business from the very beginning.

People fear change, it's just our nature. And transformation means change. AI just puts it under a spotlight. People don't fear technology itself; they fear becoming irrelevant to it. They fear being left out of the story.

Leaders today are failing when they're framing AI as an efficiency play when it's actually a transformation play.

The Sales Numbers and What They Really Meant

"Net sales in the fourth quarter were $2.54 billion; an increase of 10.3 percent compared with net sales of $2.30 billion during the same period in fiscal 2008. Fourth quarter same-store sales increased 6.9 percent compared with the same period in fiscal 2008"

—Nordstrom Q4 2009 Earnings Release, February 22, 2010

Meanwhile, the sales were blistering. Online sales were unfathomable. A technology project was headline news throughout the company.

For those final four months of 2009 we added $250M in topline sales, significant in an $8B business.

But to me the biggest impact was the business shift. There was no more Nordstrom Direct and Nordstrom brick-and-mortar. They were now intertwined. The vocabulary of company changed. We cared about cancel rate, fulfillment rates, thresholds. Technology was no longer a back-office function. It was front and center to the business. It mattered at all levels.

Perspective shift

After a decade at Nordstrom, I reached out to Mike Richardson, the CIO, for career advice. I knew I could trust him for truth, and he did not disappoint. I wanted to become a CIO. Before he advised me on how to get there, he wanted me to understand what I was saying.

"Julie, here's what you need to know. As CIO you sit at the intersection of every function yet belong fully to none. Your work is invisible until it isn't. You know more about the company's inner workings than anyone else, but you can't always share what you know. You have to make high stakes calls without complete information, and you have to get them right more often than not.

Trust is your real currency. You have to earn it from people who have no way to evaluate your work except by how you show up, how clearly you explain the unexplainable, and how quickly you solve problems they don't understand. You have to keep the future in view while the present is pulling at your sleeve. And you have to do it without flinching, because the moment you lose your composure, they'll decide the system is failing."

I asked him if I could get there from withing Nordstrom. He told me clearly, "no, you cannot. You need to go to another company to get the title."

For Your Leadership

At Nordstrom, I learned that being right means nothing if no one hears you, and that influence happens in conversations, not presentations.

Understand How Decisions Actually Get Made. My funding request was denied even though my analysis was solid. A mentor explained, "the formal meeting isn't where influence happens. It's where influence gets confirmed." I'd been playing the wrong game entirely. I thought if I prepared better, had sharper insights, people would listen. But organizations don't work that way. Influence isn't about having the right answers. It's about helping other people arrive at those answers with you. Once I learned to build conviction one conversation at a time, every funding request was approved. Not because my analysis improved. Because I finally understood how influence really worked.

Transformation Requires True Partnership: The omnichannel breakthrough happened because business and technology worked as equals—business brought customer insight, technology showed what was newly possible, and together we discovered opportunities neither could see alone. AI amplifies this need. The best strategies emerge when business problems meet AI capabilities in genuine collaboration and a discovery process where AI can inspire new business models and business insight grounds AI in real value.

Trust Is Your Real Currency. A mentor told me, "you sit at the intersection of every function yet belong fully to none. You have to earn trust from people who have no way to evaluate your work except by how you show up, how clearly you explain the unexplainable, and how quickly you solve problems they don't understand." That's true for any leader driving transformation others can't fully see.

Making the decision to leave Nordstrom after 10 years was one of the hardest things I have ever done. I loved the company and developed my foundational leadership beliefs there. The decade I spent there formed me as a technology leader. I had a front-row seat to transforming how technology could drive business results. But after a decade, I knew I needed external experience to grow my career.

I thought I was leaving to learn new skills at a new company. What I didn't realize was that the most profound transformation was already beginning. But not in a conference room, but in the streets of Addis Ababa, where a boy named Ermias would teach me lessons about connection, belonging, and leadership that no business transformation ever could.

CHAPTER 5

When Love Expands

Adoption

I spent years learning to see systems everywhere. How data flows through networks. How decisions ripple through organizations. How one change in a retail system could unlock millions in revenue. But while I was learning to see systems in technology and business, I was still blind to the system I needed to understand most: influence. I could solve complex technical problems. I didn't know how to solve the ones that involved love, fear, or trust. I understood dependencies in code, but not the dependencies we create when we commit to people. Adoption changed that. It forced me to see that the most powerful transformations don't come from optimizing processes. They come from expanding who you're willing to become for someone else.

All adoption stories are complicated, and each one is unique. There's rarely a simple narrative. More often, there's loss, love, impossible choices, and families doing the best they can in circumstances most of us will never face. What I share in this book I do with permission from Ermias, now eighteen, and our family in Ethiopia.

Opening the Door

Ever since Mason came into our lives, we had wrestled with the question of adoption. We weren't sure if our family was complete. We decided to take in exchange students in part to help answer it. What would it feel like to have three kids? How would it impact Kenzie and Mason? Can they accept another person as a sibling or are we already at our max?

First, we had Felix from Germany. He was a soccer player and had a strong business mind. He was a sponge at learning and immersed himself in every experience he could. Kenzie and Mason, then eight and six, loved him like a big brother. Suddenly Germany was a place where their brother lived, no longer some far away country. Then we had Borja from Spain. Borja was also a soccer player and the most loving, fun, caring person we could have imagined. He lit up our house and made everything feel like a celebration. Soon Spain became a place their other brother lived. The world got smaller and more relatable.

Maybe the turning point was when Mason, cheeky even at eight, asked me, "Mom, can I have a brother I can keep?"

My brother and sister-in-law had just adopted two boys from Ethiopia, and watching their family showed us it was possible. The love and connection were immediate. We knew we could do it too.

We turned in the first round of paperwork, eager to see where this journey would lead us.

From the start, we knew this was not something we could control, so we left ourselves open to be led. For us, that meant being willing to be amazed at the perfection that comes from something bigger than human plans. This openness, this act of surrender, has led me through all major changes in my life.

The adoption process moved quickly. Within months, we saw a picture and two short videos of a beautiful six-year-old boy who instantly melted our hearts. He was spunky, smart, social, and loved soccer. And *left-footed*, something that both of my brothers and my father noticed immediately. He had been in the orphanage for two years. The agency staff had fallen in love with him too. They told us, "He is special." One of the women who visited him described his sense of humor. She said that during one of her visits, he walked up and asked, "Madelline, can I be you today?" She agreed, and he sat down with her paper and pen, looked at her, and asked, "Madelline, what is your name?"

In the videos, he clapped loudly for his playmates, hogged the soccer ball, and grinned straight into the camera after kicking it hard. I showed the video to Mason. His first words were, "He kicks the ball hard. I like him."

The First Trip

When we landed in Addis Ababa for the first trip, the air felt different. Thin but warm, with the scent of spices and diesel from the streets below. The city was alive with the sounds of car horns, the calls of street vendors, and the rhythm of conversations in a language that felt like music. We drove past tin-roofed stalls, donkeys in the street, women balancing baskets on their heads, friends walking hand in hand, and clusters of children waving at our van.

I brought Mason and my best friend James. Cindy stayed home. Ethiopia treated same-sex relationships as a crime. Adoption rules closed the door on couples like us. We had to hide part of ourselves and move forward as if I were a single mother. James, Mason and I spent two weeks at the orphanage, getting to know Ermias and the community he lived in, while the legal process moved forward.

Meeting Ermias for the first time was both ordinary and extraordinary. He wore a too-small shirt with a bright pink vest over it and held a soccer ball so tightly it looked like he would not let it go. He clearly ran that orphanage. His eyes locked with mine with the confidence of someone who already knew who he was. He did not need us to define him. In that moment I understood that adoption was not about rescuing or saving. It was about joining lives, honoring what was already there, and building something together.

The orphanage was clean but crowded, filled with the sound of kids playing and the steady hum of caretakers moving from room to room. I noticed how the children looked out for one another. They shared toys, steadied each other as they climbed steps, and offered a hand without being asked. It was a kind of interdependence I had rarely seen in the United States, and it made me rethink my definition of strength.

The older kids welcomed the new arrivals by showering them with love, alleviating their fears. They ate together as one community around a small round table, meals shared from a single platter, hands reaching in together, laughter spilling out easily.

Over those two weeks, I began to see the beauty in the way community was woven into daily life. Strangers touched your arm when they spoke to you, grounding the conversation in presence. Time felt different. It was measured less by schedules and more by connection.

When it was time to leave without him—the legalities weren't complete yet—I was heartbroken. The girls at the orphanage sensed it. They gathered around me and braided my hair, an act of love to comfort me. I sat there, tears streaming down my face, while small hands worked through my hair with such tenderness.

Meeting Wude

On that first trip, I asked the orphanage director if I could meet the woman who had brought Ermias there two years earlier.

She was very solemn. Through an interpreter, she told me how much her family loved Ermias but that she had a special needs son, and the extra burden of another child was too much. She told me she didn't think he would be in an orphanage for two years, that she had searched for a good orphanage, and was so glad he now had a family.

I gave her a necklace I had worn every day. I promised her that I would take good care of him, that he would be happy and well loved. She shyly took it, we took a picture, and then she was off. But not before she invited me to her home on my next visit.

The exchange was brief, but something about it stayed with me. There was a story beneath the story, layers I couldn't yet see.

I left that day knowing that there were so many cultural queues I missed, things I didn't understand. I wanted more time to let her know me and to know her, to understand why this place that shaped my son was so special. I wanted him to know that too, yet I knew I was about to make that very difficult.

Coming Home

Three months later, after the US processed all the paperwork for Ermias' citizenship, I returned to Ethiopia. This time with Cindy, MacKenzie, and my 80-year-old loving, but still stubborn, mother. After following my dad around through baseball trades, moving 39 times in 14 years with three small children, she was no stranger to travel and certainly not one to be left behind.

This time, we could finally bring Ermias home. Cindy wouldn't miss that moment, and we were careful about perceptions.

The First Night with Ermias

That first night in the orphanage all five of us settled into a small bedroom in a guest house. We shared no common words with Ermias. We looked and sounded very different from the people he had known. In his single act of defiance, as Kenzie and I laid down in the double bed, Ermias turned his feet toward our heads. I understood, but it still hurt. This child was our family, and yet we had no way to connect with him as he stood on the edge of another round of enormous change and loss.

The Gift from Strangers

The next morning we walked the dirt streets, past two lazy donkeys and several random chickens, heading to a coffee house. We kicked rocks and tried to find connection with Ermias.

Ahead we saw a woman walking toward us, dressed in a black, head-to-toe burka. As we got near to each other, she began to sing to us, "There was a farmer who had a dog and Bingo was his name-o." Without missing a beat, Ermias replied, "B-I-N-G-O." In that instant she gave us a gift---a way to connect. We sang the song together over and over all the way to the coffee house. A point of connection. What a gift.

When we reached the busy road where the coffee house sat on the other side, I bent down and offered to carry Ermias.

He refused. He wasn't ready.

I took his hand so we could cross safely.

Once inside, Ermias was more interested in the people around us than in staying at our table. He spotted two Ethiopian sharply dressed businessmen and wandered over to them, placing his arm along the back of one man's chair. I was right there and called him back, embarrassed, but the man looked up, smiled, and said in perfect English, "It is okay." He pulled Ermias close, speaking to him in Amharic with warmth and confidence. The second man joined in, their voices steady and encouraging.

I sat down with my mom, Cindy and Kenzie at our own table.

I observed the interaction, listening without understanding, watching their kindness wash over him. They gave him their full attention. They let him play with the coins on the table. Their care was immediate and unconditional. They set aside their own conversation as if nothing mattered more than this little boy they had just met.

At the same time I was embarrassed. I wanted to show these men that I was a capable parent and could also engage with my son, who obviously wanted nothing to do with me. I called him back to us. He ignored me. Every ounce of my being wanted to control this situation, but I looked over at Cindy who gave me a calming, "let it go look" as she modeled a deep breath.

The men called me over and pulled out a chair for me. As soon as I approached, Ermias left the table. They told me, "it's okay, let him go." They asked me his story. Then my story. They wanted confirmation that he did not have a family who could care for him. That was what I believed, and I told them his story. Then they called him back to the table. He returned and stood between the two of them. They spoke to him in Amharic. All I could understand was one word: *Obama*. They then told me what they had said to him. That his life would be good, that his family loved him, and that he should love and trust us in return. They told him that maybe one day he would grow up to be Obama.

When we stepped out of the coffee house, Ermias stopped and turned toward me.

He lifted his arms to me.

I carried him back across the street, feeling the shift in that moment. For the first time, he had chosen connection over independence. From that point forward, he considered us family and never looked back.

Never, not once.

The impact those men had on our family is something they'll never know. But they taught me the hardest lessons I'm still learning. The lessons of letting go of control, learning from others, trusting, being present. Years later, I'd realize this became my most powerful leadership tool. Connection first. The rest follows.

The kindness of strangers in Ethiopia was not random. It was rooted in a deep awareness of one another that I do not often see elsewhere. People seemed to know they were part of a whole, and they acted on it. They gave time, welcome, and presence. Sometimes it was in the form of a song from a passing woman who only showed her eyes and her voice. Sometimes it was two men in a coffee house, setting aside their own conversation to encourage a boy they had just met. These moments taught me that the smallest acts can become the bridge to trust, and that bridge can change everything.

Inside the coffee house—
the two men who changed our lives

Visiting Wude's and Eyayu's Home

Before we left Ethiopia, we made that visit to Wude's home that she had invited us to on our first trip. She prepared a traditional coffee (buna ቡና) ceremony for us, roasting the beans over open coals, the smoke and coffee smells filled the small room. She introduced us to her husband, Eyayu. What I witnessed challenged every assumption I had about Ermias' early life. I saw the love of her family, the depth of connection, the way they moved together as one unit. I understood how painful it must have been for them to let go of him.

Her 19-year-old daughter, Bruktawit, spoke some English. I asked her if she missed Ermias. She looked at me with the most serious expression and said, "Yes, he is my brother." She did not mean it only in the Western sense of family by birth, but in the African sense of closeness and belonging.

In that moment, I realized adoption wasn't just about gaining a child. It was about understanding loss—the profound loss experienced by those who loved him first. Wude was very sad to give him up, but she also had to maintain the story she had told the courts, which was that he was the son of a transient neighbor. I had a feeling I wasn't getting the whole story, and it would take me years to get it.

Meeting our Ethiopian family for the first time.
My mother on the left, Ermias in grey striped shirt in front.

Landing Home

When we boarded the plane to bring Ermias home, I carried more than a new role as a mother. I carried a new lens that would stay with me every time I stepped into a room as a leader.

I had gone to Ethiopia thinking I knew what I was doing. I was the capable executive, the one who could manage complex situations, the problem-solver who always had a plan. And in that coffee house, two strangers showed me that sometimes the most powerful thing you can do is get out of the way and let others lead. A woman on the street who only showed me her eyes saw exactly what I needed.

I had spent years building walls to protect myself at work. Performing competence. Managing perception. Making sure no one saw the parts of me that felt vulnerable or uncertain. And here was this six-year-old boy teaching me that real connection only happens when you stop performing and start being present.

I had believed that leadership meant having the answers. Ethiopia taught me that leadership often means asking better questions and trusting people to surprise you with their wisdom.

These weren't abstract lessons I filed away for later. They were visceral, immediate truths that changed how I showed up from that point forward. I couldn't go back to leading the way I had before. Something had cracked open, and there was no putting it back together.

I didn't know it then, but everything I would build in the years to come, every team I would lead, every transformation I would guide, every hard conversation I would navigate, it all traced back to those days in Ethiopia. To two businessmen who showed me what generosity looks like. To a woman who loved my son first and taught me about loss. To a boy who chose when to trust and reminded me that belonging can't be forced.

The foundation I thought I was building for Ermias turned out to be the foundation he was building for me.

Years later, when I would stand in front of my team and admit I didn't have all the answers, I would think about that coffee house. Years later when I was building global teams, so much of my Ethiopian lessons came back to me.

When I would push back against "onshore/offshore" language that dehumanized global teams, I would think about Ermias and the power of words to include or exclude.

But in 2013, I didn't know any of that yet. I just knew that I was flying home with a son who had chosen to trust me, carrying lessons I hadn't known I needed to learn.

And I was still hiding. Still performing. Still convinced that the masks I wore at work were protecting me rather than limiting me.

It would take a few more years, and a conversation over beer in a Red Robin, before those lessons from Ethiopia would fully land.

But the seeds were planted. And they were already starting to grow.

Settling In

Within months of coming home, Ermias was developing English and stopped speaking Amharic. We tried interpreters, Ethiopian restaurants, phone calls with his orphanage friends. He would understand everything but only respond in English. We hired one more interpreter, the eleventh. He was a kind Ethiopian businessman in a suit. He spoke to all of us in English, but as soon as he starting speaking Amharic, Ermias walked away from him. The man tried hard, but Ermias would only give him eyebrow rises and nods of his head. Then finally, Ermias told him, "I. Speak. English." The man turned to me and said, "your son has spoken."

I was heartbroken. On one hand, it showed he felt safe, bonded, fully part of our family. On the other hand, I worried we were losing something precious. I wanted him to know his full story, his traditions, his identity, his first language. That tension—between belonging and maintaining identity—became one of the most important learnings I carried forward.

Years later when we built our India Tech Hub, so much of my Ethiopian lessons came back to me. Learn and love the culture, see what's different without judgment, but curiosity. Integrate by building on top of the differences, don't erase them. Respect each other for their humanity and let yourself be transformed. The goal isn't assimilation—it's authentic belonging.

For Your Leadership

Ethiopia taught me that real understanding happens on the ground, not in presentations, and that transformation requires stepping into unfamiliar territory with curiosity.

Ground Your Strategy in Real Context: Walking the streets of Addis Ababa and sitting in Wude and Eyayu's home taught me what adoption really meant. You can't lead transformation from conference rooms. Go where the work happens. Watch how people solve problems. Understand the context your algorithms will enter. When you build from ground-level understanding, you create technology that fits how people work.

Trust Lets You Move Through Uncertainty: Sitting in that coffee house, everything felt uncertain. But we could move forward because we built trust. Transformation works the same way. You won't have perfect information. You can't eliminate every risk. But when you build relationships strong enough to navigate uncertainty together, you can make progress even when the path isn't clear.

Curiosity Opens Possibilities Control Never Could: Transformation happens when you step into unfamiliar territory with curiosity instead of control. When you're genuinely curious about what your teams see that you don't, you unlock insights no amount of authority can force. Curiosity is your competitive advantage.

Connection Across Difference Is Your Advantage: Our family expanded across continents because we chose to see beyond what made us different. Teams need the same capabilities to connect across functions and disciplines. The leaders who can translate and build understanding across these differences create organizational magic.

Ethiopia expanded what family meant to us. It taught me that the most important transformations happen when you step into unfamiliar territory with curiosity instead of control. That lesson that real understanding requires going where the work happens would become essential in ways I couldn't have predicted. Because while I was learning to navigate one kind of transformation personally, I was about to walk into another professionally. REI would test everything I thought I knew about leading through crisis.

CHAPTER 6

REI: Climbing out of Crisis

After a decade at Nordstrom, I'd reached a natural stopping point. Mike Richardson had been right. I needed experience outside of the company to grow. Career and family were both undergoing major changes. Our family was now complete with MacKenzie (12), Mason (10) and Ermias (7). Ermias was learning English at record speed.

The lessons from Ethiopia were still settling in when REI approached me. The conditions seemed ideal for someone who loves transformation: new CEO, new agenda, lagging technology. I thought that taking a role in a smaller company, a co-op even, would give me the chance to spend more time with my family.

The job reported to the "head of technology" and covered all applications across the enterprise. So, broader in remit, but shallower in scale. I liked my boss. We had dined together a few times and talked about the path ahead. I was excited.

Within three months of me starting, she left, and I was in an interim position leading the entire technology organization.

Suddenly I wasn't just responsible for applications. I now had cyber security, network, infrastructure, data and analytics. Everything. It was a much bigger challenge, and I loved it. A steeper learning curve, exactly what I wanted.

But I'd already scheduled a two-week European vacation. My family, two close friends, meeting up with our "European sons," the exchange students who'd lived with us and become family. The trip was locked in. Non-negotiable.

Then I learned my first board meeting was scheduled for the Monday I got back.

The Vacation I Didn't Take

I considered cancelling. My team needed direction. Systems were breaking. The business was a mess. This was objectively terrible timing for the interim leader to disappear to Europe.

But if I cancelled, what message would that send? That work always comes first. That family is negotiable. That the interim leader doesn't take vacation because she's too busy proving she deserves the permanent job.

So I went.

And I worked the entire time.

Coffee shops in France. An AirBnB in Tuscany. Train stations between cities. My family would explore during the day and I'd carve out evenings to build the board presentation. But it wasn't just a presentation. I was trying to prove I belonged in this role.

I had my team send everything. Every initiative in flight, every risk, every business outcome. Retail systems, rei.com, merchandising, supply chain, cyber security, network, infrastructure, omnichannel, corporate systems, data and analytics. I read it all. Memorized it. Sent questions back and forth. Built my own deck showing current work, a draft roadmap, risks from my early vantage point.

I wasn't present on that vacation. Not really. My body was in Europe but my brain was at REI headquarters, trying to absorb everything fast enough to look like I knew what I was doing.

I landed Saturday. My boss sent a note, "can we meet before the board meeting at 7AM? I'd like to discuss your role."

Discuss my role?

Panic hit immediately. They'd done the search. They'd found someone. I was about to get demoted back to the smaller role I'd originally signed up for.

Jet lag mixed with dread. I didn't sleep Saturday night. Or Sunday night. By Monday morning I was a wreck. I was foggy, exhausted, and barely functional.

I walked into that 7AM meeting carrying resignation.

He said, "Congratulations. You're our new CIO."

What.

I had not prepared for this scenario. My head was already in a fog and now it was spinning. Brand new CIO. First board meeting in nine hours. I grabbed my notes and studied more. By 4PM when I was scheduled to present, I was barely conscious. Excited about the title, sure. But absolutely not at my best.

I stood in front of the board and adrenaline kicked in. I got through the presentation. I was so prepared it wasn't hard. I knew every initiative, every number, every risk.

Then the questions started.

"Julie, given the amount of work ahead, what would it take to go faster?"

I froze. I hadn't thought about that. I fumbled around in my mental database looking for the answer instead of just having a conversation.

Next question, "what technologies should we be leaning into that we're not?"

Not on my quiz notes.

I gave answers. They weren't terrible. But they weren't good either. I was performing confidence instead of thinking clearly.

On the same day I received my first CIO title, I gave them reason to question whether they'd made the right choice.

What I Learned From Bombing

I walked out of that boardroom knowing I'd failed.

Not because I didn't know the material. I knew it cold. But I'd prepared for a test instead of a conversation. I'd memorized facts instead of building judgment. And I'd ignored every signal my body was sending me because I thought powering through made me stronger.

It didn't. It made me worse.

I learned that you can't perform your way to confidence, and that lesson hit hard. Real confidence isn't having all the answers memorized. It's being able to think clearly when you don't.

The next board meeting, I came prepared but I also came honest. When someone asked a question I hadn't anticipated, instead of scrambling through my mental files, I said, "That's a great question. Let me think through the trade-offs..."

I paused. Thought out loud. Acknowledged what I didn't know and what I'd need to explore. I treated them as partners thinking through complexity, not examiners grading my performance.

The energy shifted immediately. They leaned in. Asked follow-up questions. Offered their perspectives. We had an actual strategic conversation instead of a Q&A where I was frantically trying to prove I belonged.

That's when I understood that Boards don't need the CIO who memorized every answer. They need the CIO who can make sense of complexity in real time and make sound decisions with incomplete information.

One week into my new role as CIO, I got to test whether I'd actually learned that lesson.

When Everything Breaks

It was 9:01 AM.

I had been in the CIO role and for under a month. The title still had stickers on it.

Phones in every store stopped ringing. REI.com showed a 500 error. Every register froze. Wi-Fi and email were down throughout the headquarters campus in Kent, Washington, a suburb of Seattle. Revenue was halted simultaneously, system wide, and without warning.

I stayed calm, while also knowing the urgency. This wasn't my first outage, and I knew we would get through it. But I was also hyper-aware of how the team was responding to this critical incident.

The networking team was scrambling to diagnose the issue. Nothing was documented, but the tenured leaders knew how to navigate. They narrowed it to one server, and a few hours later, found the culprit. It was a firmware patch on a *single server* sitting under a table about fifty feet from my office. HP had issued a fix weeks earlier. We hadn't installed it.

This outage revealed the cost of treating technology as a mere back-office function. A single server sitting under a table took down stores, registers, the website, and HQ simultaneously because there was no redundancy, no monitoring to catch issues before they became catastrophic, and no process to track and deploy critical patches HP had issued weeks earlier. The team relied on institutional knowledge rather than documentation, and the architecture allowed one component failure to cascade across every system with no isolation or graceful degradation. It was a textbook example of accumulated technical debt in a company that hadn't invested in resilience because nobody thought technology was critical enough to warrant it, until everything stopped at once.

That outage was devastating in the moment, but it gave me exactly what I needed to make the case for rebuilding.

Using Crisis to Build the Case

This outage gave me something I couldn't have engineered: proof that our infrastructure was held together with duct tape and hope. The board needed to see it, and now they had.

I'd learned something from bombing at my first board meeting. They didn't need me to perform technical mastery. They needed me to frame the problem so they could make informed decisions. Different job entirely.

So instead of burying them in technical details or defending our choices, I told them a story.

"Six weeks ago, a single server sitting under a table fifty feet from my office took down every store, every register, and our entire website. For hours, we couldn't take a single dollar of revenue. Not online. Not in stores. Nowhere."

I paused.

"The server failed because we hadn't installed a firmware patch HP issued weeks earlier. We didn't install it because we had no process to track critical updates. We didn't catch it early because we had no monitoring. And when it failed, it took everything down because we had no redundancy."

One board member asked, "How long until something like this happens again?"

"I can't tell you it won't happen tomorrow. We're running on borrowed time. Technical debt accumulated over years. Single points of failure everywhere. No visibility into what's breaking until revenue stops."

Another board member leaned forward. "What would it take to fix this?"

"We need to rebuild our infrastructure from the ground up. Move to modern cloud architecture. Build monitoring so we see problems before they become crises. Create redundancy so one failure doesn't cascade. It's an eight-figure investment. The alternative is accepting that we'll keep having outages we can't predict or prevent."

They approved the investment. Now I had funding, but I needed people. And I needed HP to treat us like we mattered.

A few weeks later, I was invited to an executive roundtable with HP's CEO, Meg Whitman. She wanted to listen to customers. When she asked "How can HP be a better partner?" I raised my hand.

I told her I was the CIO at REI. I told her about the outage: the server under the table, the firmware patch we'd never installed, the complete system failure. Then I asked: "How can a $2 billion co-op get the help of a $110 billion public company like HP?"

Her answer: "I think you just did."

Whitman tagged REI as a special case study. She assigned senior architects to co-design the new environment and put her own name on the escalation path. A month later, my infrastructure director told me his HP rep was fielding so many internal check-ins that he'd started asking for less attention. Good problem to have.

With funding secured, HP committed, and a clear mandate to rebuild, I thought the next challenge was talent.

It wasn't. The challenge was me.

The Beer Conversation

I had just gotten the CIO title. More responsibility, more visibility, more pressure. I reached out to Mike Richardson, the former CIO at Nordstrom, to pick his brain about the role.

We met at the Red Robin in Issaquah. He ordered a light beer. I scanned the wine list and settled for a glass at $5.99. Neither of us was much of a drinker, but the noise and chaos of Red Robin felt right for the conversation.

I didn't wait for the drinks. I launched into my ninety-day plan consisting of org charts, communication strategy, stakeholder mapping, risk mitigation. I had it all figured out.

Mike let me finish. Then he asked, "How are you going to get your team to follow you if they don't even know who you are?"

I took a sip of wine, wishing I'd splurged for the $7.99 option. I started to answer, stopped, paused, started again. Why was this such a hard question?

"I'm showing them through the strategy. Through delivering results. Through…"

He shook his head. "They know what you do. They don't know who you are. You have a wall up so high, nobody can see you. And until they do, they're never going to trust where you're taking them."

The word that came into my head was vulnerability. That's what he was asking for. My defenses went up immediately. Vulnerability? In a place where being a woman already made me visible? Where being gay could still cost me credibility, maybe even my job? Where showing any uncertainty might confirm what people already suspected—that I didn't really belong?

"I know," he said quietly. "I know what you think it will cost you. But consider what it's already costing you."

I didn't deflect. I paused. I listened. I let it sink in.

It was the first time I imagined a different way.

What made that conversation possible was the foundation Mike had already built. I'd watched him lead at Nordstrom with incredible EQ. He read people. He paid attention to what they did over what they said. He led by knowing his team deeply.

The first time I had a 1:1 with him, I felt like he was peering into my soul. He didn't want to know what I'd done. He wanted to know who I was and what made me tick. Terrifying, because my masks weren't working. I didn't want him to see any part of me except my competence and expertise. I started talking faster, filling the dead air he was giving me—the exact space I needed to hang myself.

That discomfort turned into trust. I could always go to Mike for honesty. When I asked questions others might deflect, Mike gave me the truth. Not comfortable truths. Real ones.

If someone else had pushed me toward vulnerability, I would have dismissed it as naive. But Mike had earned the right to challenge me. I trusted him. He'd seen through my performance before I even knew I was performing.

Trust. That word was new to me in a work setting.

That conversation cracked something open.

The First Test

I didn't transform overnight. Mike's question sat with me for weeks. How was I going to get my team to follow me if they don't even know who I am?

I knew I couldn't do it on my own. I couldn't even fully picture what "authentic leadership" looked like, let alone embody it. I remembered an executive coach who had known me for years, someone who had seen my strengths and my edges. So, I reached out and asked him to help me.

He asked me the questions I didn't know how to ask myself. *When do you find yourself managing perception instead of being present? What are you protecting by hiding?*

The answers were uncomfortable. I was protecting an image of competence that required perfection. I was protecting myself from the judgment I assumed would come if people knew I was gay. I was protecting a career I'd built by making myself acceptable.

But I was also realizing the cost. My team didn't trust me because they couldn't see me. I couldn't build the kind of culture I wanted with psychological safety, truth-telling, and authentic connection while I was still performing a version of myself I thought they needed.

The first shift was small. A Monday morning team meeting, the kind where people share weekend updates. Someone asked what I'd done.

For years, I'd edited Cindy out of every story. "Friends and I went hiking." "I was working on the house." Careful pronouns. Vague references. Never a "we" that could reveal too much.

This time, I didn't edit.

"Cindy and I took the kids to the park," I said. Then kept talking like it was the most natural thing in the world.

The room didn't gasp. Lightning didn't strike. One person asked about my kids. Another mentioned their own weekend. The meeting continued.

But something in me had shifted. I'd stopped protecting the image and started showing up as myself.

The next week, I did it again. Mentioned Cindy in passing. Talked about our family without editing who we were. Each time, it got a little easier. Each time, I felt a little less exhausted by the performance I'd been maintaining for decades.

What surprised me wasn't that people accepted it. It was that most of them had already known or assumed and were just waiting for me to stop hiding. The masks I'd thought were protecting me had just been keeping everyone at a distance.

Within a few months, something had changed. Not just in how I showed up, but in how my team showed up. People started being more honest in meetings. More willing to admit when they were stuck. More willing to challenge ideas instead of staying quiet.

I'd spent years thinking my job was to have all the answers and never show weakness. In fact, what I was learning was that the strongest thing I could do was stop pretending.

Hiring for a Mission

The city of Kent, Washington, wasn't Seattle. It was thirty minutes south without traffic, an hour with it. By reputation, it wasn't a cool place to be a technologist. *At all.* Amazon was hiring the same talent in South Lake Union with modern buildings, free food, and a great vibe. The co-op had a reputation as where "technology careers go to die," as someone told me when I was leaving Nordstrom.

But REI's mission became our advantage. We attracted technologists who wanted to do great work for a cause they believed in. Leaders who led with values. Outdoor junkies, community volunteers, co-op members who cared about the mission as much as the stack. Engineers who designed for the future, comfortable killing their own ideas before failure did it for them. Architects who could diagram a spine-and-leaf network and still explain why inventory checks had to clear in under 300 milliseconds.

The ones who were chasing purpose, not prestige. That filter made everything else work. New hires integrated fast because the mission filtered them in the first place. Stand-ups opened with mountain bike trail reports as often as they did burn-down charts. Code quality reviews included questions like, "What's the sustainability impact if we ship this?" The work mattered because it served something larger than the sprint goal.

In short order we outgrew Kent. We had engineers using windowsills as desks. We moved twenty minutes north to Bellevue, which opened us up to another pool of talent. But the bigger win was cultural. People came because they believed in what REI stood for.

Building a mission-driven team was one thing. Becoming the kind of leader who could truly lead them was another.

Mike had challenged me to let my team know who I was. The coaching had helped me see my masks. But knowing you need to be authentic and really doing it in front of people are very different things.

That's when the invitation arrived.

When Vulnerability Became Influence

When a local Women in Tech group asked me to speak, I said yes. Then immediately regretted it.

I chose the topic "authenticity" because I knew it would require me to be, well, authentic. That was what I was working on. But choosing the topic and standing on that stage were very different things.

Before I go there, let's talk about authenticity as a leadership principle. A lot of us are talking about this, and some are even saying authenticity can have negative consequences. For me, authenticity is not about just trusting your gut and speaking from the heart. It certainly does not mean to stop learning because you're just "being authentic." Authenticity paired with awareness, responsibility and curiosity is where the magic happens. I can stay anchored in my values while I adjust how I lead as the stakes rise, teams grow, or contexts shift.

Real authenticity isn't "this is who I am, take it or leave it."

Real authenticity is "this is who I am becoming, and I am willing to grow in public."

I didn't have that clear of an opinion, however, the day of the event. My knees wobbled. My mouth went dry. I was used to speaking on stage, presenting business cases and technology strategies. But not this. Not vulnerability in front of a crowd.

My mother sat in the first row. She was always my biggest fan, no matter what I was talking about, especially if I was talking about myself. But that made it harder somehow. I was about to say things out loud I'd spent decades hiding.

I stood there questioning why I'd agreed to this. But by then, I was committed.

So I talked about what it was like being a woman in tech. How hard it had been. How I'd built masks to protect myself. How being gay had caused me to live in fear. How I'd made myself smaller to fit in, and what that had cost me. The things I had learned.

I said the things I'd never said publicly before.

When I finished, I wasn't sure what would happen. Maybe I'd miscalculated. Maybe I'd shared too much. Maybe everyone would politely clap and move on to the networking reception.

That's not what happened.

Women flooded toward me. They wanted to talk. They had stories to share. One after another, they started with the same phrase, "I've never told anyone this before, but..."

We stood there as the room emptied around us. The organizers started cleaning up. Eventually, they had to turn the lights off in the building.

That's when I understood what had just happened. My vulnerability had given them permission. By being honest about my struggle, I'd created space for theirs.

I'd spent years trying to build influence through competence. Through having the right answers. Through proving I deserved to be in the room. And in one hour of vulnerability, I'd created more connection than years of performance ever had.

That was the shift. Influence wasn't about being the smartest or the most polished. It was about being honest enough that others could be honest too.

That Women in Tech speech changed something fundamental. I'd discovered that vulnerability creates connection. Now I had to learn what to do with that connection.

The test came sooner than I expected.

Using the Platform

I started paying attention to the impact of my words, not just the message I wanted to deliver. When I spoke, I paid attention to both what I wanted to say and what my audience needed to hear. Sometimes they were different. I could feel the impact as I was speaking, which gave more weight and importance to every word.

But more than that, I started looking for opportunities to create space for others. To stand up when it mattered.

At REI, the executive team debated whether to publicly support marriage equality. I sat at the table with the CEO, Jerry Stritzke at the time, to my right, and the rest of the leadership team around the same table. The concern wasn't whether it was the right thing to do. The concern was backlash. Whether this was really an issue the company needed to be vocal about.

I took a breath.

This was 2015. I wasn't the only one sending messages to the CEO and executive team, but I changed the dialog at the table in that moment.

"This issue matters to me and it matters to our employees," I said. "Our employees need to see that we take a stand for what we believe is right. I've lived in hiding for many years without feeling the protection of companies I work for. Knowing where REI stands gives all our employees the certainty to know that their jobs are safe and their company truly supports them."

The team paused. I felt heard.

We supported marriage equality publicly and I'm proud that the company continues to take a stand for their beliefs.

That moment wasn't about me looking brave or progressive. It was about using the platform I'd built to create certainty for people who didn't have access to that table. Every LGBTQ+ employee in every REI store across the country would know where their company stood. I was no longer hiding; I was using my influence to stand up for the community I was a part of.

Technology as Enabler

Nine months after that server under the table brought us down, we launched Store Fulfillment. This time, learning from Nordstrom, we partnered with stores from day one. Members could shop the full network inventory, orders went straight to the right store, and associates had the tools and training they needed. Same square footage, higher turn, happier members.

By year two, the technology organization had shifted from reactive firefighters to a team on the front end of the strategy opening new revenue streams: used gear resale, geo-fenced event pushes, and deeper loyalty integrations. The Board thought of us as a back-office function; instead asked what we could do with more money.

REI doesn't exist so people can shop. It exists so people can get outside. To show its values it closes its doors on Black Friday to encourage its members to go outside instead of participating in the consumerism of that day. Every build decision had to shorten the distance between a member's intention and their first step on the trail.

The tech wasn't the story; what it enabled was. Members got outside faster, the co-op grew, and the team saw their code echoed in bikes on trails, tents under stars, and jackets traded for a second life. That alignment between circuitry and cause became the blueprint I would later carry to lululemon.

A New Challenge

After two and a half years at REI, I'd proven what I came to prove. We'd rebuilt infrastructure from crisis. We'd turned technology from a cost center into a growth engine. We'd created a culture where mission attracted talent that built capabilities our competitors couldn't match.

But more than that, I'd learned something about myself. I could lead without the masks. The Women in Tech speech, the marriage equality conversation, the way I built the team…I was finally becoming the leader Mike had seen in me years earlier. Someone who could navigate systems without disappearing into them. Someone who could use influence in service of something bigger.

I could feel the ceiling approaching. The next level of growth that meant scale, global complexity, and strategic influence wouldn't happen inside the co-op.

Which is why, when lululemon called, I was ready.

The kids, now 14, 12 and 10, were thrilled about the brand switch. "lululemon!?! How do you even *know* about lululemon?" That's right, kids, I'm doing this so you have more relevance in your social circles.

REI taught me that catastrophe creates permission to rebuild. At lululemon, I would learn what happens when you actually use that permission—not just to fix what's broken, but to build something that scales. The company was growing at rocket speed, and everything I'd learned about foundations, culture, and authenticity was about to be tested at a level I'd never experienced. This time, there would be no practice run.

For Your Leadership

You Can't Perform Your Way to Confidence. I bombed my first board meeting because I'd memorized facts instead of building judgment. I treated it like a test instead of a conversation. Real confidence isn't having all the answers. It's being able to think clearly when you don't. Preparation means building judgment, not memorizing facts.

The Wall That Protects You Is the Wall That Limits You. Mike Richardson asked me a question that changed everything: "How are you going to get your team to follow you if they don't even know who you are?" I had built a wall so high that nobody could see me. I thought it was protecting me. It was isolating me. The masks we wear to feel safe are the same masks that prevent real connection. People follow leaders they trust, and they trust leaders they can see.

Crisis Creates Permission to Transform. That server under the table exposed what happens when you don't own what matters. The outage was devastating, but it gave me exactly what I needed to make the case for rebuilding. Don't waste a crisis. Use it to tell the truth about what's broken and build the case for what needs to change.

Sustainable Leadership Requires Real Boundaries. I took that European vacation but didn't really take it. I was physically in Tuscany while mentally grinding through documentation. By the time I stood in front of the board, I was too exhausted to think clearly. The leaders who survive transformation know when to step away and actually step away.

REI taught me how to build from a breaking point and use crisis as a moment to highlight the need for transformation. But the real test was waiting. At lululemon, I'd face the challenge of building culture at scale, taking everything I'd learned and proving it could work when the stakes were highest and the timeline was shortest.

Part 2: Transformation

The foundation was built. I had perspective. I'd learned to speak my truths, to build trust, and to show up authentically. I'd learned what influence really meant, that it was about relationships or power and the difference between the two. I'd also learned that you don't need antagonist relationships in order to succeed. I'd learned partnerships get the best outcomes when all parties want the same goal.

But all of that was preparation for what came next.

At lululemon, I'd face the challenge of transforming a $2B company into a $10B global powerhouse in less than a decade. Building culture at scale. Leading through pandemic. Expanding across continents. The question wasn't whether I knew the principles. It was whether those principles could hold under real pressure.

At the time, though, I had no idea how much this theory would be tested. At scale. At speed. And with a fast-growing company very dependent on beating those tests. It was where the theory met reality. Where the capabilities we've been building get tested. Where we'd discover if what works in one building can work across the world. Where we'd learn that culture is more than something that you create and are done. Instead, it's the work itself. It needs to be operationalized, scaled and protected. Every single day.

lululemon: Where Culture Met Scale

The foundation I'd built over decades—systems thinking from behind home plate, self-worth learned through walking away from disrespect, the ability to see potential in others that they couldn't yet see in themselves—was about to be tested at lululemon. And that test began immediately.

A Foundational Decision

"E-commerce comps rose 60 percent on a constant-dollar basis... we streamlined checkout in Q1 and are bringing data-driven, automated tools to mobile, search, browse, email and post-purchase experiences."

—lululemon earnings call, Q1 2018

In my first week, after we had recovered from the 20-hour outage, I learned we were committed to launching a new North American website before the holiday season, just five months away. I started on a Wednesday, the day the consultants onboarded me with a deep dive into the work they were in. Thursday I went to Portland. Friday I went to San Francisco. By Friday evening, I knew we had a problem.

The internal teams weren't working together. The San Francisco team owned the core platform but only communicated via tickets. The Portland team, working with the mobile team in Vancouver, had built a promising cloud-based prototype as a way around the commerce platform, yet lacked a clear integration path. And the consultants? They were speaking directly to executives with the promise of a pre-holiday launch, highlighting the biggest risk of execution.

Each group thought the others were either uninformed or incapable. There was distrust, frustration, and no alignment.

On Saturday I knew I needed help, fast. I called a fellow soccer mom. I knew she was a program manager that was in between jobs. I admitted I couldn't pay her immediately, or even get her on the books, but promised I would. Luckily, she said yes.

I called an all-day meeting for that next Monday, day four for me, in Seattle. I brought all key leaders together—Portland, San Francisco, Vancouver, the contract program manager and the consultants—and laid it all out. We weren't working as a team, and it was those of us in the room who had to figure it out.

The energy in that room reflected everything I was learning about lululemon: bold, possibility-oriented, but also intensely impatient with anything that felt like delay. This wasn't a company that would wait for perfect information or consensus-driven decisions. The founder-speed conviction I'd felt from my first interview was still alive in the organization, even with new leadership. People expected decisions, not discussions.

It was quickly apparent that the front-end tech stack was the point of contention. The consultants had built a custom solution that their architect believed was the optimal approach for our requirements. However, our internal teams had significant concerns about the architecture's long-term scalability and maintainability. The disagreement wasn't about competence—it was about fundamentally different philosophies around build versus buy, and whether we should optimize for speed to market or sustainable foundation.

The consultants' approach prioritized getting something launched quickly with proven, proprietary software. Our internal teams were concerned about inheriting a system they hadn't designed and didn't fully understand, especially given our aggressive growth projections.

As the discussion unfolded, it became clear this wasn't just a technical decision. It was a strategic choice about ownership, capability building, and long-term control over our platform.

I asked the consultants to step out while I talked with my team privately. That's when the real concerns surfaced. They didn't just disagree with the technical approach— they didn't want to inherit and maintain a system they hadn't helped design. They were worried about being held accountable for something they couldn't fully control or modify.

When I brought everyone back together, I realized this decision would define more than just our technology stack. It would signal whether we were building internal capability or remaining dependent on external expertise.

Choosing the internal build wasn't the safest option. At REI, I'd learned that performing confidence gets exposed under pressure. I'd bombed my first board meeting by trying to memorize every answer instead of building real judgment. That lesson stayed with me. This wasn't about whether consultants or internal teams were 'better.' It was about whether we would own our outcomes or outsource them. I'd rather have a team learning out loud than consultants performing expertise we couldn't maintain. And it was the one that could turn disconnected teams into a unified force. It wasn't just about launching a site. It was about proving what was possible when people.

Later that night, I spoke with the internal leaders. What I observed was different. They had *conviction*. They believed their approach could work. More importantly, they wanted to *own* it.

The next morning I stood in front of the same room and said, "we're going with the internal build. Not because it's perfect, but because we're going to own our future. And we're going to do it together."

Then I made the accountability explicit. I said, "we own this outcome now. If we miss the deadline, there's no vendor to blame. If the site crashes, it's on us. I'm betting on your conviction, but you need to deliver." Trust doesn't mean lowering standards. Rather, it means being crystal clear about what success looks like.

Some executives, including my boss, were skeptical. The consultants were shocked. But the team? They showed up. They worked around the clock. They problem-solved. They collaborated. They rebuilt trust because now, it was theirs.

You can't outsource belief. You can't vendor your way to ownership. Transformation doesn't come from the cleanest pitch. It comes from the people who are ready to do the work and feel the outcome. When teams build together, they begin to believe in each other. That belief is the engine, and that's what drove my decision.

I chose to bet on our team's conviction. A strategy from consultants can be smart, but it's vulnerable if no one inside believes in it.

You can absolutely buy great tools. Sometimes that's the smartest move. But you can't buy the energy and pride that comes when a team builds or shapes something themselves. The trap isn't choosing a vendor. It's assuming the vendor replaces your responsibility to build capability.

The October deadline arrived like a freight train. We had to decide whether to go live with a solution we had limited infrastructure to fully test. No performance tests, no A/B tests, no rollback plan. Just go or no-go.

Standing in that WeWork conference room, I looked around at the faces of people who'd worked weekends, who'd rebuilt systems from scratch, who'd turned disconnected teams into a unified force. The choice wasn't really about risk management. It was about values.

"We're going," I heard myself say. "And we're going to watch every minute of it together."

It worked. The site held. Sales soared. Conversion improved. But the most important result wasn't in the numbers. It was in the room. That one decision, the risky one, sparked belief. Not just in the platform, but in each other. In what we were capable of. I had trusted them. And now, they trusted me.

The numbers were impressive. Forty-two percent digital growth, eighteen percent overall revenue increase, conversion improvements across the board[3]. But the real victory was watching a fragmented team become a unified force.

Growing the Leadership Team

Before we could modernize systems or deliver new capabilities, we had to build leadership that could unify culture, set vision, and execute against it. The fragmentation was more human than technical.

I needed leaders who could learn fast and lead forward, people who could hold a vision, ask hard questions, and partner with the business in shaping what came next.

My first hires after Janine, my Executive Assistant, were people I'd worked with before. Not because I wanted comfort, but because I needed to have trust with people immediately. In stable times, you can build trust slowly. I didn't have that luxury. Fires were daily, the organization skeptical, and transformation work urgent.

I looked for leaders who could make calls with incomplete information, keep teams moving when nothing's certain, and defend the vision when I wasn't there to do it myself. That takes tested trust.

Venki Krishababu and I met at Nordstrom, where we'd known each other mostly by reputation. Vaidynathan Seshan (most people call him "Seshan") came most recently from REI, where I'd seen his character tested under pressure and watched him lead two companies through data transformations. Both had the technical depth and leadership presence we desperately needed. But more than that, they knew how to build real partnership, the kind where you challenge each other's thinking and get to better answers together.

I chose leaders I knew would put team success over personal credit, surface problems early, and extend trust because they had already earned mine. Familiarity wasn't the credential; their ability to extend trust quickly was.

But I also learned the risk of creating an "inner circle." When new leaders only trust familiar allies, organizations split into insiders and everyone else. Trust fractures along those lines and rarely heals.

My hires weren't my people. They were leaders for the whole organization. Their job wasn't to make me successful but to make the organization successful. They had to earn trust with teams who didn't know them, prove their value, and show they were invested in one team, not one person.

The Pull Became Magnetic

Once I got strong technical leaders in place, people like Venki and Seshan who brought both deep expertise and authentic leadership, something shifted. lululemon, which had been virtually unknown in Seattle's competitive tech scene, became a destination. The talent we attracted started attracting more talent.

lululemon's headquarters was in Vancouver, but we needed to grow in two directions at once. Vancouver's technology talent market was too small to support the scale we were building toward, so I created a technology hub in Seattle alongside the technology teams in Vancouver, where we could tap into the deep bench of engineers, architects, and product managers coming out of Amazon, Microsoft, and the region's startup scene.

We quickly outgrew our cramped WeWork space. We were adding engineers, architects, and product managers faster than facilities could keep up.

The momentum was real because the team was real. Over time, I had leaders who could amplify vision, move fast without losing people, and carry weight with me, and sometimes for me, so the organization could accelerate while staying connected.

Confronting What Was Broken

But growth was exposing cracks in the foundation. By summer, I was learning the full extent of what we'd inherited.

Two major technology initiatives were described as "in progress" when I arrived. First was a new point-of-sale (POS) system to go in every store and the second was a merchandise planning platform. Both were labeled as strategic priorities. Both projects were officially "on track."

But I'd learned not to trust "on track" until I could see the track myself. So I went directly to each leader and asked them to show me the systems.

The truth came out in careful conversations: there wasn't much activity on POS. The merchandise planning "co-development" with a vendor was a PowerPoint deck dressed up as a software platform. No real system, no integration plan, no substance underneath the polished presentations. Millions of dollars invested in each, nothing of substance.

I wasn't the technical risk that shocked me. I'd seen failed projects before. It was the silence. People knew the truth, but no one had said it out loud.

I took it to the COO. Laid it out clearly: we were chasing sunk costs on two dead initiatives. Meanwhile, stores were dealing with failing point-of-sale systems and the business was running critical merchandising operations on 72 interconnected spreadsheets that could break with a single wrong formula.

The irony wasn't lost on me. I was hired to transform technology, but my first major act was stopping technology initiatives. Sometimes before you can build what's needed, you have to stop pretending you're building it.

The organizational issues ran deeper than failed projects. Our success was making the structural problems impossible to ignore.

Location Matters: The Consolidation Decision

The website launch had shown me what was possible when we worked as one team. But it also revealed what wasn't sustainable about our current structure.

When major decisions needed to be made, we flew the Portland, San Francisco, and Seattle teams to Vancouver. For real collaboration, people started bypassing formal channels entirely. We were succeeding despite our structure, not because of it.

I had to face the truth: We were a small team trying to build something big, scattered across four cities with no reliable way to work together in real time. This was 2017. Video conferencing existed, but we hadn't built the infrastructure or culture to make it work. Conference calls with terrible audio where half the room couldn't hear. Email chains that took days to resolve what should have been ten-minute conversations.

You can't build trust when you can't read someone's face. Can't pop by their desk to work through a problem. Can't grab coffee after a tough meeting. We were too small and too new as a team to operate remotely. We needed relationships, not just org charts. And relationships require presence.

So I made the call I knew would be painful but necessary: consolidate from four locations down to two. Close San Francisco, offering everyone the option to relocate to

Seattle or Vancouver. Portland would follow later, once we'd proven the model worked.

It wasn't about cutting costs or asserting control. It was about admitting what the coordination overhead had taught me: you can't build trust across distances when you're still building it within teams. We needed to be together. To make fast decisions without scheduling across time zones. To build the kind of working relationships that let us move at the speed the business needed.

Seattle gave us access to the technical talent we couldn't find in Vancouver. Vancouver kept us connected to headquarters and the culture that made lululemon unique. Three hours apart by car. Close enough for real presence, shared culture, and faster decisions.

I didn't send emails or delegate the conversations. My leadership team and I flew to both San Francisco and Portland. I owned the decision completely and explained exactly why we were making it.

I didn't sugarcoat it or hide behind business euphemisms. Four locations had created coordination costs we couldn't afford and trust deficits we couldn't overcome.

More than half chose to relocate. This said as much about their belief in where we were heading as it did about their individual circumstances. The others, we helped transition with respect and support.

When made with clarity, transparency, and complete ownership, hard decisions can deepen trust. People can handle difficult truths. What they can't handle is being managed around them.

As we consolidated and the team began working more closely together, I also focused on building the operational discipline that would make our speed sustainable. One early example was the daily Sales reports. The numbers that executives used to understand business performance were inconsistent, and we learned about system failures from guests or business users before our technology team knew anything was wrong.

I put one leader in charge and made them accountable for daily sales reports running every single day, no matter what it took. Not accountable for the upstream systems, they didn't own those. Not accountable for the data accuracy issues, those belonged to other teams. Accountable for reports going out, every day, without fail, with communication to all affected users.

This was about the philosophy of ownership I'd brought with me, not about blame. Instead of asking "whose fault is this?" I asked "who will make sure this works?" Systems should notify us when something breaks, not the people using them. Every morning those reports went out, we were telling the organization: when Technology commits to something, you can count on it.

By fall, we had consolidated locations, killed dead projects, built operational discipline, and assembled a leadership team that could move fast together. We'd proven we could deliver the North American website. We had the foundation to scale.

And then came the real test.

The Olympic Test: Korea and Partnership

My second month, while I was buried in hiring, relationship-building, and system stabilization, I learned that the Asia-Pacific (APAC) team wanted to launch a Korea website for the Winter Olympics. Bold digital entry into a fast-growing market, timed to a global event that would put lululemon in front of millions of potential new guests.

The business logic was compelling. The technical reality was sobering. Our architecture couldn't handle regional scaling. But business leaders weren't asking for technical transformation—they were asking for launch.

So I called Ken, the General Manager of the APAC region. He was everything I'd heard and more. Dynamic, smart, relentless. The kind of executive who got things done whether systems were ready or not. The conversation was direct from the first minute. He didn't trust IT. Didn't trust me. Told me so without hesitation.

I promised I was in the process of hiring a new Digital leader, and once they were on board, we would fly to Hong Kong to work through this together. We would be there within a month.

Not fast enough. Ken wanted me on a plane the next day. I asked him to please let me help, but I needed time.

The next day, I learned through a partner in legal that Ken had signed a contract with a large vendor to build Korea's website. He'd done exactly what his track record suggested he would do. He found a way forward that didn't depend on trusting anyone else.

I had a choice that would define my relationship with the business for years to come. I could let him proceed knowing it would likely fail and use that failure as proof he needed me, or I could intervene immediately and risk destroying a relationship that had barely started.

So I called the vendor directly. Introduced myself as the new CIO and asked them to walk me through the project they'd just contracted to deliver. I wanted to understand scope, architecture, localization, fulfillment, guest experience, and payment integration.

What I discovered confirmed my worst fears. There wasn't a real project plan. No localization strategy for Korean language, currency, or tax requirements. No integration plan for inventory, fulfillment, or financial systems.

I told them calmly but directly: "I'm the new CIO with a charter to build international technology capability. You have a choice: tear up this contract and be part of a larger RFP process for all international websites where we can do this right or honor this contract and let Korea be the last lululemon business you do on my watch."

Fortunately for both companies, they chose to tear up the contract.

Ken was furious, and I didn't blame him. But I also knew that contract would have set us up for failure on a global stage.

I told him the truth: "Ken, this contract won't deliver what you need for Korea to succeed. Not because I want control over your region, but because I care about getting this right for your business. You may not know me yet, but I keep my word. I'll deliver this with you, and I'll be on a plane to Hong Kong in two weeks to prove it."

Two weeks later, my newly hired digital leader's first day on the job was Saturday, and we were on a plane to Hong Kong. What followed was relentless execution. The team worked around the clock, building integrations around missing pieces, patching together systems never meant to work together across continents. But it was live before the Olympics started.

Speed came at a price I should name explicitly. The team worked brutal hours. We accumulated technical debt that took years to pay down. And I had to have hard conversations afterward: "We proved we can move fast. Now we prove we can move sustainably. The next launch won't be heroic effort. It'll be repeatable process." Some people who thrived in crisis mode struggled with that transition. That's when I learned that different phases of transformation need different capabilities.

The Olympic success opened doors, but it also led to one of the most challenging confrontations of my leadership journey.

The Breaking Point

After the Korea launch, I brought the full truth to the COO, my boss at the time. Yes, we'd hit the deadline and had a website live for the Olympics. Yes, we'd proven we could deliver under impossible pressure, again. But we hadn't built a scalable solution. We'd built a one-off that worked for Korea but couldn't be replicated efficiently for other markets without significant rework.

I expected that nuance to matter. I was wrong.

The conversation erupted in the lobby of a Vancouver leadership conference. He wanted to hear about victory, not complexity. He wanted confirmation that technology could deliver miracles on demand, not explanations about what miracles actually cost.

I don't raise my voice, ever. But that day I did, matching his energy level. Not to win an argument, but to be heard. To show that I cared deeply about getting this right and was willing to fight for what mattered.

Standing in that lobby, I realized I had a choice that would define not just my career, but who I was as a leader: stay true to my integrity or stay in the role. If telling the truth cost me my job, so be it.

That night in my hotel room, I drafted my resignation. Not in anger or wounded pride, but in clarity about what I could and couldn't accept. I couldn't lead transformation while being asked to stay silent about what needed transforming.

I put the email in my "24-hour bin," the rule my Dad had taught me for big decisions made in emotional moments. But as I fell asleep that night, I knew something had shifted. I'd found my line in the sand.

The next day, we sat down again, just the two of us, away from conference crowds and lobby acoustics. Having written that resignation letter, having found my absolute limits, I felt different. Calmer. More centered.

I told him directly that I needed his support to back the work we were doing. I did not need him to shield me from difficult conversations or unrealistic expectations, but to help me navigate them productively.

The breakthrough came when he realized I wasn't trying to win an argument or protect my territory. I was trying to build a future we could both believe in. We had the same goal of making lululemon capable of becoming what it was meant to be.

That conversation didn't end with resolution or celebration. It ended with understanding, and a small step toward trust. Harder won but more durable than easy agreement.

Two years later, Ken bought me a nice bottle of sake to celebrate what we'd built together. Not the Korea launch. The partnership. The digital progress we had made across Asia that had grown his business. The trust that came from choosing truth over convenience, from keeping promises even when they were hard. That bottle of sake meant more than any performance review ever could.

A Seat at the Table

A year in, I was invited to join the Executive Leadership Team, reporting directly to the CEO, Calvin McDonald.

It was quiet but transformative. I wasn't just head of technology anymore, keeping systems running. I had a seat where company strategy was shaped.

Years earlier, Jamie Nordstrom had crystallized how retail viewed technology. I'd asked him about real career opportunities for someone like me. His answer was direct: "Julie, there are two functions that matter in retail. Stores and merchandising. We think of IT like the boiler room. If something breaks, we go in, turn the lights on, rattle things around, turn the lights back off, and go back to business. If you want a career in retail, you need to be a store manager or a merchant."

I had no interest in either.

That conversation became my North Star. Not because I accepted his assessment, but because I was determined to prove it wrong.

Now, sitting at the Executive Leadership table, technology had moved from the boiler room to the boardroom.

The change wasn't about title or org chart. It was about access to context. I was in the room where priorities were set, trade-offs debated, direction determined. I could shape business strategy with technical reality baked in from the start.

Before, I'd receive filtered requests: "the business needs this by Q3." Now I heard the conversation behind it. Why Q3 mattered. What market pressure was driving it. What we'd sacrifice to deliver it. Whether timing was actually negotiable or just wishful thinking.

It was like someone had finally taken the blinders off.

That context changed everything. During discussions about international expansion, the conversation shifted from "We're entering China, make it work" to "What markets can our architecture support now? Where do we need foundation first? What's the sequence that actually makes sense?"

I wasn't answering technical questions after strategy was set. I was helping shape which markets we entered, how fast we could scale, what competitive advantages we could actually sustain.

When we debated guest experience improvements, I could say "That requires rebuilding our entire order management system. Six months minimum. But here's what we could do in six weeks that gets you 80% of the impact." We could make real trade-offs in real time instead of technology becoming the bottleneck that slowed everything down.

When merchandising wanted to test new product categories, I could explain what our systems could handle versus what would break. We made smarter bets because technical constraints were part of the strategy conversation, not discovered after decisions were made.

My team felt the shift immediately. They finally had representation at the highest level. More importantly, they started getting strategic direction instead of conflicting priorities filtered through three layers.

Decision speed changed overnight. Technology proposals no longer cycled through approval chains. We could debate, decide, and move in one conversation.

Business leaders started bringing us in earlier. They stopped viewing technology as the group that said "no" and started seeing us as partners who could help figure out what was possible. The questions changed from "can you build this?" to "what should we build together?"

When the head of Digital sat in product discussions, people listened differently because they knew he had direct access to strategic decisions. When our architect proposed changes for international expansion, business leaders engaged seriously because technology wasn't back office anymore. We were strategic.

Jamie and I would later laugh about his boiler room comment. It was time-stamped to a world before technology was woven into company strategy. But I never forgot it. It reminded me how far we'd come.

For Your Leadership

Those first months at lululemon taught me that trust is earned in production and that killing failed projects openly builds stronger culture than hiding them.

Invest in Leaders Who Believe Before It's Proven: I chose people who believed in what we could build together, not just people with perfect résumés. Look for people who believe in your mission, who are willing to learn alongside you, who care more about outcomes than credentials. The person who's curious, committed, and aligned with your vision will deliver more value than the supposed expert chasing the next title.

Credibility Comes From Shipping, Not Talking: The Website launch and the Olympic Test in Korea proved we could deliver under pressure. Credibility doesn't come from pilot presentations. It comes from deploying systems that solves real problems and showing up when things break to fix them.

Kill Failed Projects Openly: When I confronted dead projects publicly, it signaled that honesty mattered more than performance. You will have failed pilots, models that don't work, vendors who overpromised. Kill them openly. Explain what you learned. Show that failure is acceptable when you face it honestly. The organizations that hide failures perpetuate them. The ones that confront failures openly build cultures where people surface problems early.

Trust Scales Only With Intentional Action: That Vancouver confrontation taught me that good intentions aren't enough. People judge you on what they experience. When you say "responsible AI" matters, people watch which projects you fund. When you claim "human-centered approach," they notice who's in the room. Make every decision visible evidence of what actually matters to you.

Building the lululemon foundation taught me that trust and truth-telling could transform technology organizations. But as the company scaled from $2B toward $10B, I faced a new question. *Could culture scale the same way technology did?* Could psychological safety work across continents? Could authenticity translate across languages? The next phase would test whether what worked in one building could work in dozens of countries.

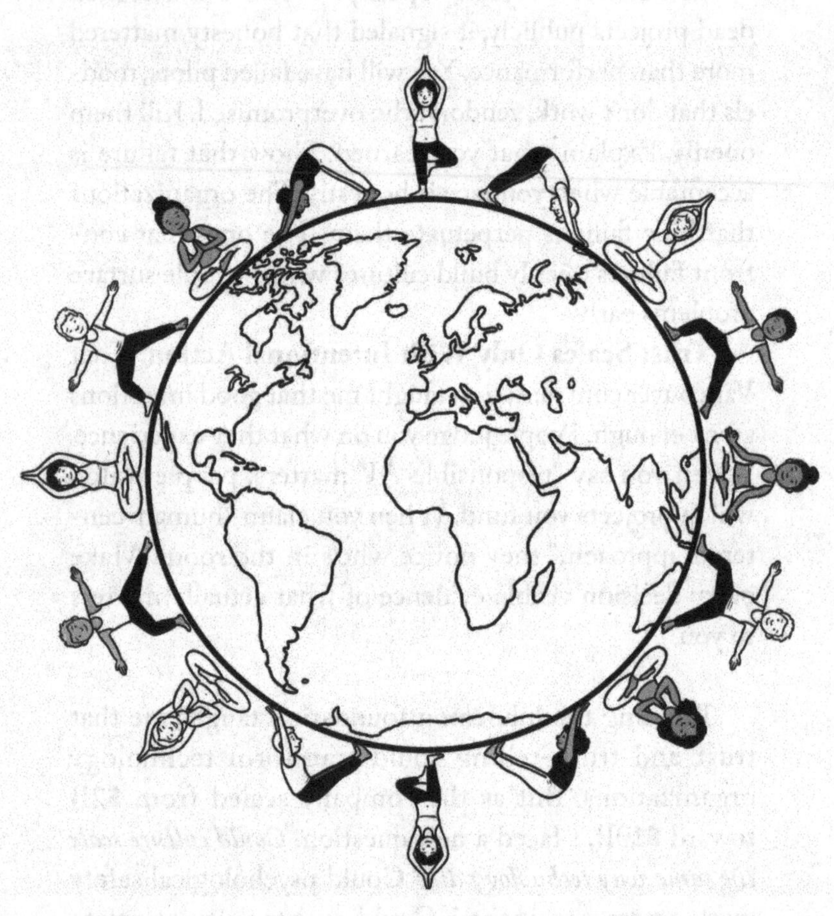

CHAPTER 8

Building One Team

The Vancouver leadership team sat across from me, arms crossed. We were six months into the brand-new office space in Seattle, and the rumors had been circulating for weeks. People were afraid that Seattle was taking over. The real leaders were all in Seattle now. Vancouver was being sidelined.

"Is it true?" one of them finally asked. "Are you replacing us?"

I could see it in their faces. I saw the fear of becoming secondary. That "one team, one culture" was just code for "Seattle calls the shots." They'd watched me hire Venki and Seshan, both now based in Seattle. They'd seen the rapid growth south of the border while Vancouver stayed relatively stable. The math, from their perspective, told a clear story.

"No," I said. "That's not what's happening. But I understand why you'd think it is."

I couldn't argue with the pattern. From their perspective, the evidence was damning. I was based in Seattle. My closest leadership hires were in Seattle. The new roles I was hiring for had Seattle in the job description.

"Watch what I do, not just what I say," I told them. "Judge me on decisions, not proximity."

I meant it. But I also knew that words (even honest ones) wouldn't rebuild trust. Only actions would. Only showing up, again and again, in ways that proved Vancouver mattered just as much as Seattle.

That conversation taught me that you can have the right principles, the right frameworks, the right intentions, and still watch trust fracture if you're not paying attention to what people are actually experiencing. This is critical in scaling culture.

Building culture at one site is hard. Scaling it across borders while keeping it authentic? That requires relentless intentionality.

Reducing Fragmentation

We had no engineers and no architects, a history of plug it in and let it run. The mindset was "capital is free" with support costs hidden in future projects. We were held hostage by vendor contracts signed with no real choice.

But underneath the chaos, I saw high commitment, creativity, and pockets of real talent. The problem wasn't the people. It was the lack of structure, strategy, and shared accountability.

After consolidating from four scattered offices to two strategic locations, I had geography working for us. But proximity doesn't create belonging. Leaders still operated in silos. Smart people lacked strategic direction. Individual contributors worked around each other instead of with each other.

This wasn't just a project management problem. It was systems architecture that reflected deeper organizational fragmentation. We called it "the hairball," a mess of connections where touching one system broke three others, or twenty. We were running 200+ concurrent projects with minimal oversight.

The real bottleneck was human, not technical. We had single points of failure throughout the organization. We had people who held knowledge so tightly that the entire system depended on them. That dependency had to be broken, even when it meant difficult decisions about who stayed and who left.

Behind the scenes, I was cognizant that building a Seattle-based senior leadership team risked creating a culture divided by location. But I needed a team with deep retail experience who had strong ownership and would bring the rest along. I knew I needed to address this tension openly.

We said no to projects that didn't matter. That gave us room to fix what did. The architecture was necessary, but belief made it possible. You can't transform systems people don't believe in.

Building Leadership

Looking at this chaos, I could see the root cause was leadership. We had capable people operating as individual contributors rather than as leaders who could unify teams and own outcomes. The fragmentation would continue until someone took responsibility for connecting the pieces.

That's when I knew I needed to start with the leaders who could influence everyone else. Leaders with a strong sense of curiosity, ownership, and charisma.

When I think about the people who shaped lululemon's early transformation, Venki Krishnababu stands out not just for what he accomplished, but for how he accomplished it. Venki joined to lead store technologies, infrastructure, and architecture—critical areas that were struggling under the weight of rapid growth and would later become our CTO.

At Nordstrom, we'd known each other only by reputation. Then when I landed at lululemon, Venki told me he wanted to work with me. This was back when our Seattle team consisted of Janine, myself, and a wooden logo in a WeWork. When Venki came in for our conversations, his intensity and passion were unmistakable, but I hadn't worked closely enough with him to know his full capabilities.

I called someone who knew Venki well. He told me, "If I were going to start a new company, he would be one of my first hires." A builder, an innovator, and a trusted ally—that was all I needed to hear.

I hired Venki knowing he had the technical depth and leadership presence we desperately needed. What I didn't fully anticipate was how naturally we would work together, how our different strengths would complement each other, and how that partnership would become a model for the culture we were building.

From the start, Venki didn't sit behind a desk. He spent time in the stores, shadowing sales people (AKA educators), asking questions, and listening carefully. He wasn't satisfied with surface answers. He wanted to understand the frustrations of a slow register, the ripple effect of a dropped network, the pressure of a long line on a Saturday morning. He led by showing up where the work happened.

Venki earned trust by being clear and direct. He set a vision that reliability was not negotiable, then broke it into steps the team could own. He brought discipline without stripping away creativity, insisting on strong processes but leaving room for ideas to bubble up. His teams knew where they stood, and they knew he had their back.

Within months, registers hummed reliably, networks stayed stable, and our own help desk was supporting Educators directly. Walking into stores became a source of pride rather than anxiety. That shift wasn't just about systems—it was about how Venki led.

But the real magic happened in how we worked together. We built something on high trust and constant communication. Our long walks became legendary, meandering conversations where we'd hash through whatever was occupying our minds. We didn't just divide and conquer; we led as partners, each bringing our strengths to challenges that required both of us.

If Venki was about building solid foundations and reliable systems, Vaidyanathan Seshan (Seshan) was about seeing around corners. He joined to lead analytics and data science, and later would run all of Product, Planning and International technology, and he brought the ability to turn data into decisions before problems became crises.

I'd hired Seshan at REI and watched him transform the company from one that looked backward at reports to one that looked forward with predictive analytics. I tried to hire someone else but when the role opened a year after I arrived and Seshan was looking for a job, I knew it was meant to be. He understood kind of culture we were trying to build.

What struck me about Seshan (and what I'd seen at REI) was his charismatic energy. He had this way of making everyone in the room feel like any problem was solvable. When he talked about what data could reveal, you could see possibilities you hadn't imagined before. That optimism was infectious and genuinely useful when you're trying to transform an organization that's been stuck in old patterns.

Seshan worked relentlessly. He'd paint ambitious pictures of what was possible, then put in the hours to make them real. When timelines needed adjusting or models required refinement, he'd come back and say, "Here's what we learned, here's what we're adjusting." That combination of enthusiasm and follow-through kept people believing even when the path got harder than expected.

What made Seshan's optimism credible was how deeply he knew his people. He didn't just know the names of everyone on his team—he knew their strengths, their growth edges, what motivated them. He could tell you which analyst was brilliant at pattern recognition but needed help communicating to executives, which data scientist was ready for a stretch assignment, who was struggling personally and needed some slack.

That wasn't performative. Seshan genuinely cared about developing people. He'd spend hours coaching someone through a tough presentation or helping them think through a career decision. His team was fiercely loyal because they knew he saw them, not just their output.

Seshan understood that data without context is just noise. He spent time with merchandising teams, talking to store leaders, understanding the rhythms of the business. He could translate between the language of algorithms and the language of guest experience. That bridge-building made analytics accessible instead of intimidating.

What I valued most about Seshan was his strength of character. He never weaponized data to win arguments or protect territory. If his analysis showed that someone else's approach was working, he'd champion it. If his own hypothesis was wrong, he'd say so. That humility, combined with his technical depth and charismatic energy, made him someone people wanted to collaborate with rather than compete against.

Like Venki, Seshan understood that leadership wasn't about being the smartest person in the room. It was about making everyone in the room smarter. He built a team of analysts who didn't just crunch numbers but partnered with business leaders to solve real problems. He created a culture where asking good questions mattered more than having quick answers.

As Seshan grew into his role, his impact became impossible to ignore. He wasn't just delivering analytics—he was changing how the entire organization thought about data. That's when I promoted him to senior vice president. It wasn't about giving him a title. It was about recognizing that he had become the leader the business couldn't function without.

Between Venki's operational excellence and Seshan's vision, we were building the technical backbone lululemon needed. But more than that, they were modeling the kind of leadership I wanted to scale. They lived competence without arrogance, collaboration with accountability, and ambition with humanity.

The partnership between the three of us became a living example of what "one team" could look like. We challenged each other, supported each other, and trusted each other enough to disagree openly. That's what I wanted to replicate across the organization.

What Venki and I had built together—high trust, constant communication, shared leadership—became the template I wanted to scale across the entire organization.

The First Test

After that first all-hands leadership meeting, the Vancouver directors hung back. One of them pulled me aside in the hallway. "That was a good speech, Julie. But Seattle got Venki and Seshan. You're based in Seattle. The new roles you're hiring for? All Seattle job postings. So when you say 'one team,' what you mean is 'one team reporting to Seattle.'"

I couldn't argue with the pattern. From their perspective, the evidence was clear.

"You're right," I said. "The optics are terrible. But watch what I do, not just what I say. Judge me on decisions, not proximity."

I meant it. But I also knew that words wouldn't rebuild trust. Only actions would. They would later see that we built leadership across both cities, promoted tenured employees into key positions, and grew the Vancouver team to be bigger than Seattle. But that would take time. For now, we were operating on borrowed trust.

Building the Muscle

Six months later, we were presenting detailed execution plans to executives. We had technology leaders assigned to each business segment. We had clear owners for every milestone, performance metrics tracking conversion improvements, and roadmaps stretching months ahead. The contrast with our early chaos was stark.

A year later, we gathered again for another leadership offsite. This time, we went deeper. We committed to meeting regularly, holding quarterly offsites in Seattle or Vancouver, and building trust through empathy and vulnerability. We worked within the culture of lululemon and instituted "Practice of Leadership" sessions focused on "Leading Self" through "Personal Responsibility," "Possibility," "Integrity," and "Mindful Communication."

But the real shift wasn't in our meeting cadence or our vocabulary. It was in how people saw their jobs. We were building something together.

Creating Safety

Clear expectations weren't enough. Leaders could understand their new roles intellectually but still fall back on old patterns. Protecting territory, guarding information, playing it safe. To truly transform how they worked together, I needed to address the fear underneath.

At one offsite, I asked three directors to share their personal leadership journeys. I'd spoken with each of them beforehand, one-on-one, to explain what I saw in them that could help the group. What they brought into the room surprised me. Not polished leadership narratives, but honest, grounded stories about struggle and growth. They weren't there to perform. They were there to connect.

Tears came quickly. Then laughter. Walls came down.

Suddenly there weren't separate cities anymore. We were one team. Jokes started flying. The tension that had been suffocating us lifted.

What I watched happen in that room, Google's research later confirmed. In their Project Aristotle study of 180 teams, they found the answer wasn't talent or resources. It was psychological safety, a concept Harvard professor Amy Edmondson developed to describe what happens when people believe they won't be punished or humiliated for speaking up with ideas, questions, concerns, or mistakes. When people feel safe, they learn faster, collaborate better, deliver stronger outcomes.

Safety Doesn't Mean Comfortable

Building psychological safety meant people could admit when they were stuck. But it didn't mean staying stuck was acceptable.

I had directors who said, "I don't know how to lead at this scale."

My response sounded familiar, even as I said it, "good, that's honest. Now what are you going to do about it?"

My dad had asked me the same question years earlier, when I called him heartbroken after those two managers resigned. He'd listened to everything—the exit interviews, the feedback that I was disconnected, my confusion about how I'd gotten it so wrong. Then he said, "Now you know. What are you going to do about it?"

Not sympathy. Not excuses. Just clarity about ownership.

Some directors invested in their development and grew. They got coaches, took on stretch assignments, practiced new approaches.

Others kept expecting the problem to solve itself.

Those conversations got harder. We were raising the bar. Every week they didn't adapt, the gap got wider.

"Six months ago you said you needed support. We gave you coaching, stretch assignments, peer learning. You're still not stepping up. That's a choice. And it has consequences."

That's what real safety looks like. Honest conversations about what's working and what isn't. Clear expectations. Support to grow. And accountability when people choose not to.

Safety doesn't mean protecting people from hard truths. It means being honest about what's changing while helping people see where they fit in what's next.

Making Culture Visible

Frameworks and strategies set direction, but culture only becomes real when it shows up in rituals people recognize and repeat. Small, visible practices that turn values into muscle memory.

At lululemon, one of the first rituals we created was the leadership offsite. Not meetings, but gatherings where we brought vulnerability, honesty, and clarity into the room. They became cultural markers. A signal that leadership meant accountability and truth-telling, not title or tenure. Directors shared their own stories of leadership lessons, personal growth, and hardships. It was always optional. But as trust built, sharing vulnerably became common.

We built rituals for personal storytelling too. When new leaders joined, we invited them to share a six-minute Pecha Kucha introduction. It was a simple format. Twenty slides, twenty seconds each. Mine included hashtags like #FamilyLove, #Awkward, #NoTouchie. Funny, imperfect, very real. The point wasn't polish. It was presence. These sessions gave teams permission to bring their whole selves, not just their resumes.

Another ritual was Practice of Leadership sessions, structured around leading self, leading others, and leading enterprise. Not lectures. Conversations where leaders practiced integrity, mindful communication, and accountability together. Over time, the phrase "Practice of Leadership" became shorthand for how we expected leaders to show up every day.

And there was one ritual that ran through the entire company, which was starting every meeting with a clearing. A brief personal check-in, sometimes just a sentence, where each person named what they were bringing into the room so they could be fully present. "I'm distracted because my child is sick." "I'm excited because I just finished a marathon." Clearings weren't therapy sessions. They were acknowledgments. Reminders that people are human first, roles second. They created space for empathy before we got into agendas.

Over time, the clearing became a cultural symbol. We each cared enough to see the person, not just the job.

Making Culture Operational

Culture had to be built into our operating system, not treated as something we'd lean on only when things went wrong. It needed to be what made everything else possible.

So we wove it into the fabric of how we worked.

Interview guides didn't just test for technical skill. They tested for ownership. Could you admit when you didn't know something? They tested for integrity. Would you surface a problem even when it made you look bad? They tested for collaboration. Did you see others' success as connected to your own?

Performance reviews measured what we actually valued. "Builds trust" sat right next to "delivers results." "Coaches candidly" mattered as much as hitting milestones. Because we'd learned the hard way that leaders who delivered outcomes while destroying teams weren't successful, they were liabilities on a delayed fuse.

Promotions required evidence of enterprise outcomes, not just individual wins. You couldn't climb by making yourself indispensable. You climbed by making others capable.

And recognition? We told stories about the moments that defined us and matched our values. Not the big launches or the revenue numbers, but the engineer who admitted a critical bug before anyone else found it. The director who pushed back on unrealistic timelines to protect the team. The manager who chose the harder right over the easier wrong, even when it cost them personally.

This was how culture outlived any one leader. It wasn't dependent on Venki and me walking the halls or hosting offsites. It was embedded in the very mechanics of how people entered, grew, and were measured. It became the system itself.

The Promotions that Proved It

By 2019 we had promoted four Vancouver-based directors into Vice Presidents. They had tenure, business relationships, and proximity to the business.

I promoted the Vancouver director who'd confronted me two years earlier about being "sidelined."

A Seattle leader pushed back in the leadership meeting. He said, "are you sure? That's a global role. Shouldn't it be based here where you are?"

"No," I said. "It should be based where the best leader is. And he's in Vancouver."

That promotion signaled that location didn't determine leadership potential. Capability did.

The metrics told part of the story. Between 2018 and 2019, our employee engagement scores climbed from 72 to 88 percent. Turnover dropped by half. Time-to-hire for critical roles decreased by 40 percent as our reputation spread and talent sought us out instead of us chasing them.

But the real measure wasn't in the numbers. It was in how people showed up.

Leaders who used to wait for permission started making calls. Teams that used to protect information started sharing it freely. Directors who used to compete for resources started problem-solving together. The culture we'd been building intentionally was becoming the default way we worked.

Full Circle

Three years after that confrontation in the hallway where I was asked "Are you replacing us?" I was in Vancouver for a leadership offsite. The same director who'd asked that question approached me during a break.

"Remember when I accused you of making Vancouver second-tier?"

I nodded.

"I was wrong. Not about what it looked like. You were hiring all the senior leaders in Seattle. But I was wrong about what you were building. You didn't make us Seattle's satellite. You made us equals. And then you made us hold you accountable to that."

He paused. "That's when I finally believed it. When you trusted me enough to lead without needing to be the one in control."

That conversation meant more to me than any award or recognition. Because it proved that culture doesn't scale through mandate. It scales through trust, built one decision at a time, tested and proven through actions that match words.

The foundation was set. Not because it was built perfectly, but because it was built with intention, integrity, and the stubborn belief that we could create something better together than any of us could create alone.

We had built the leadership capability. We had created psychological safety. We had made culture visible and operational. We had proven that "one team, one culture" could work across cities when you invested in making it real.

By early 2019, the culture we'd built looked solid on paper. Leaders were collaborating across cities. Psychological safety was operational. Our rituals were embedded. But there's a difference between building culture and proving it works under pressure.

The real test wasn't whether people could be vulnerable in offsites. It was whether they could hold high standards while maintaining safety.

That test came sooner than expected.

When Innovation Gets Isolated

As we scaled, the pressure to innovate faster intensified. New capabilities, new markets, new guest experiences. The demand was relentless.

The conventional wisdom said we needed to create an innovation team. Give them space separate from operational pressures. Let them explore emerging technologies without the constraints of production systems.

So we tried it. We carved out a small team, gave them freedom to think big. No quarterly delivery targets. No operational responsibilities. Just pure innovation.

Within months, the problems surfaced. The innovation team would build something promising that the operational teams couldn't integrate because they hadn't been involved. Or they'd solve problems the business had already deprioritized for legitimate reasons. The operations teams felt like innovation was being done TO them. The innovation team felt dismissed.

I realized the people running operations weren't less innovative than the "innovation team." They were just buried under operational pressure with no permission to lift their heads and think differently.

We disbanded the separate innovation team.

Instead, we started asking different questions in interviews. We designed them to reveal curiosity, ownership, collaboration style, conflict approach. Do you naturally see problems and start thinking about ways to solve them? How do you work with people who think differently than you?

Because if you hire curious problem-solvers and then bury them in process without space to think, you've wasted what made them valuable in the first place.

What truly unlocked innovation wasn't a framework. It was hiring people who cared about the problems, then creating conditions where they felt safe to experiment. That's what created ownership.

That meant protecting time for exploration. It meant celebrating smart failures as loudly as successes. It meant when someone said "What if we tried this completely different approach?" the answer was "Let's test it" not "That's not how we do things."

The breakthrough ideas didn't come from a separate innovation lab. They came from the engineer who'd been managing our deployment process for two years and finally had permission to completely reimagine it. From the product manager who knew our guests better than anyone and was trusted to try something unconventional. From the data scientist who sat with the business team long enough to understand what problem needed solving.

When people own the problem, they innovate on the solution. You can't hand someone else's brilliant idea to a team and expect the same commitment you get when they discover it themselves.

Innovation stopped being a separate function. It became what naturally happened when curious people had trust, time, and permission to solve problems differently.

I'm watching companies make this mistake with AI right now. They create "AI Centers of Excellence" separate from operations, staffed with specialists. But AI isn't something specialists do while everyone else keeps running the business. It's something curious problem-solvers use when they have the space and support to ask "Could this tool help me solve this problem better?"

The companies that will win aren't the ones with the fanciest innovation labs. They're the ones who hire curious people, build trust, and give them permission to reimagine how the work gets done.

Good innovation doesn't happen in isolation. It happens when you trust the people doing the work to find better ways to do it.

When Technical Decisions Test Culture

When we made the decision to sunset ATG, the platform running our e-commerce, we had engineers whose identity was wrapped with the technology.

One of them came to my office. "I'm an ATG engineer. This is what I've built my career on. Now you're walking away from that. What does this mean for me?"

I understood what he was really asking. It wasn't about the technology. It was about whether he still had value.

"Technologies come and go," I told him. "That's the job. ATG was the right platform when we chose it. Now it's not. What makes you valuable isn't that you know ATG. It's that you know how to learn. That's what great technologists do. They evolve."

We offered training. Time to learn the new stack. Support through the transition. Some engineers leaned in. They saw it as a chance to grow, to build something stronger. Others couldn't let go of what they'd mastered.

Not everyone made the journey. And that was hard. But the ones who did became better, more adaptable engineers.

We understood this wasn't just a technical decision. When Venki talked to the team about platform modernization, he didn't call it a stack replacement. He framed it as strengthening our foundation for growth. Building something bigger than any single tool.

That mattered. When people understood why it was important to the business and what it meant for them, the resistance shifted. It wasn't comfortable. But our culture could handle difficult conversations. People didn't need to agree with every decision, but they needed to understand the reasoning and feel respected in the process.

High Standards, Safe Spaces

Psychological safety without standards is just comfort. Standards without safety create fear. You need both.

I wanted teams where people could admit when they were stuck, but staying stuck wasn't an option. Feedback happened immediately. I told my team, "Feedback isn't like wine or cheese. It doesn't get better sitting on the shelf." Pulling someone aside after a meeting, "Can I share what I just observed?" is kinder than saving it for review season.

The foundation was technical skill and relentless curiosity. When someone stopped learning, they didn't just stall themselves. They stalled everyone.

Leaders weren't exempt. They didn't all need to code, but they needed to understand the work well enough to ask the right questions.

Sometimes maintaining excellence meant helping people find better fits. Early in my career, I hired someone eager to break into technology. Smart, articulate, seemed like he'd fit the culture. But he couldn't keep up technically. I gave clear feedback, but I also told him there were better jobs out there for him, ones that would take advantage of his strengths and not frustrate him. When I eventually let him go, the conversation wasn't a surprise. Months later, he sent a thank-you card. The clarity helped him find a role that fit and had a future for him.

Keeping someone in a job they can't perform isn't kindness. It's unfair to the high performers, to the team's mission, and to that person. Everyone sees the bar. Setting it clearly and holding it consistently lets everyone succeed.

We didn't celebrate hiring women, LGBTQ+, or people with disabilities. We celebrated that they built world-class systems. Performance kept doors open. When someone wasn't meeting the bar, we said, "We hired you for your potential. This work isn't meeting it. What's the gap?"

A male leader once told me, "When my team was all men, everyone shouted. Now that half are women, everyone listens. It's made me a better husband. I see my wife differently."

Real inclusion means holding everyone to the same standard.

I could have a conversation where I said, "I care about your growth and I believe in your capability" and also "this work isn't meeting our standards and here's what needs to change." Those weren't contradictory messages. They are the same message. *I respect you enough to tell you the truth.*

Ready for What's Next

By early 2020, we had something that worked. The culture could handle tension. People trusted each other. Innovation happened where the work was. Standards stayed high while people felt safe.

And then in March 2020, everything stopped.

For Your Leadership

Trust Is Built by Closing Gaps, Not Ignoring Them. When Vancouver leaders asked if we were replacing them, I couldn't dismiss their concern. I had to acknowledge what they were seeing, then prove through action that it wasn't the whole story. Gaps between what leaders say and what teams experience will always exist. The question is whether you close them or let them fester.

Rituals Beat Rhetoric. We didn't talk about psychological safety. We built it into clearings at the start of every meeting, vulnerability in leadership offsites, Pecha Kucha introductions that showed the person behind the title. Culture becomes real when it shows up in practices people repeat without being told.

Innovation Dies in Isolation. Our separate innovation team failed because they were building for people who hadn't been part of the design. The same thing happens with AI. The companies winning aren't the ones with fancy innovation labs. They're the ones embedding capability where the work actually happens.

Hold Standards and Safety Simultaneously. Psychological safety doesn't mean comfort. It means people can admit when they're stuck, and then you ask, "what are you going to do about it?" Leaders who lean too far toward comfort avoid hard conversations. Leaders who lean too far toward standards create fear. You need both hands.

Promotions Signal What You Actually Value. When we promoted the Vancouver director who'd confronted me, it showed that capability matters more than proximity. Every promotion, every recognition, every hiring decision tells people what you really believe. Make sure the signals match the words.

By 2020, we'd built the foundation systematically and tested it under business pressure. The culture had proven it could handle tension without breaking. Now it would face the ultimate test. COVID-19 would do more than disrupt operations. It would test our culture, our systems, and our identity as a brand—and show us whether what we'd built could hold when everything else fell apart.

CHAPTER 9

The Real Stress Test

In early 2020, I believed we were ready for anything.

I was wrong about what "anything" meant.

I remember hearing that COVID had hit nursing homes in Kirkland, just miles from our office. Many companies were sending people home. I was worried about everyone on my team who rode the bus, who would be at risk on public transportation.

Vancouver wasn't yet impacted. When I talked to the leaders there about shutting down the Seattle office, they asked me to wait. The stores hadn't closed yet. They were dealing with their own decisions, their own uncertainties. It felt premature to them.

One of my leaders told me I was overreacting, that we should wait.

But I'd seen what they'd seen too—at our leadership conference when our China team couldn't attend because of COVID. Everyone in China locked inside their homes because the risk was real. And now it was in my neighborhood.

I walked around the office telling people they needed to go home. People were busy working on deadlines. They weren't ready to pack up for the day, let alone for what would turn out to be over a year.

Nobody grabbed much. We all thought we'd be back in a few weeks.

After everyone was out, I grabbed my laptop, looked around the empty office, and wondered how long it would be until we returned.

The next day, the stores in Seattle closed too.

Omnichannel on the Fly: Trust Enables Speed

In that one moment, the technology organization didn't miss a beat. While guest calls continued to come in but offices were closed, the teams immediately rolled out a way to take calls safely from the educator's home. Sales shifted to the e-commerce channel instantly, and the Digital teams ensured the systems had capacity. Stores shut down, but as they gradually reopened, the Retail technology teams had put in new technology that made shopping during a pandemic possible.

With physical retail paused, we had to act fast to unlock other routes to serve our guests. lululemon had long believed in an omni approach, but COVID forced that strategy to move from aspiration to execution overnight.

What made this possible wasn't just our technology. It was our culture of ownership and cross-functional collaboration.

Enhanced Store Fulfillment

Stores that couldn't physically serve walk-in customers became shipping hubs overnight. Our inventory tracking turned retail locations into fulfillment nodes. Inventory that had been trapped on shelves suddenly became productive. This required rapid systems alignment across teams, but we reopened stores to ship product within days.

While competitors scrambled to build basic omnichannel capabilities from scratch, we were scaling what we'd already built. Across retail, companies were shutting down services due to technical limitations or launching manual processes that couldn't scale. The industry was learning the hard way that true omnichannel required years of integrated system building, not quick patches. Our real-time inventory system, integrated architecture, and years of preparation meant we could respond to crisis by expanding existing capabilities. When customers experienced inconsistent service elsewhere, they came to us as the reliable option.

The real unlock wasn't technical. It was cultural. Our directors had learned to collaborate across domains rather than optimize for their individual areas. Store operations trusted tech to build systems that worked. Tech trusted operations to execute without perfect documentation. Everyone understood the "why" behind the urgency, which eliminated lengthy approval processes. Speed came from years of building that trust, not from moving faster in the moment.

Curbside Pickup

We built curbside pickup from scratch, shifting from in-store picking, in record time. Working across business, operations and technology teams we allowed guests to place orders online, then check in via text or app when they arrived. Educators were given tablets with fulfillment software and trained remotely. It wasn't flawless, but it worked, and guests appreciated the flexibility.

The sustainable solutions principle we had established meant we built these capabilities to last, even under pressure. Our guests, who were shopping from their couches, had access to nearly all the inventory sitting in stores.

We weren't just patching together a temporary workaround. We were creating new business capabilities that would serve us long after the pandemic ended.

Digital Educator Program

We launched our Digital Educator initiative, reimagining the guest experience as a one-on-one digital connection. In a matter of weeks, we had trained our sales representatives, what we call educators, from around the world on how to host virtual styling sessions, respond to online chats, and serve guests through video calls. These educators closed over $5 million in sales through virtual interactions that year alone.

This program worked because our culture already valued human connection. We weren't trying to replace authentic interaction with technology. We were using technology to preserve and extend it.

These weren't just stopgaps. They became permanent capabilities. And in many ways, they redefined how we thought about retail. But none of it would have been possible without the foundation of trust, ownership, and collaboration we had spent years building.

Systems Thinking Pays Off for Platforms

The foundation that carried us through COVID-19 wasn't built during the pandemic. It was built years earlier, when we made bets that seemed smart but not yet urgent.

Before I ever arrived, we'd invested in RFID technology—tiny radio-receiver tags that let us track every single product in real time. At the time, it felt like good inventory management. During lockdown, it became our lifeline. When stores shut down overnight, we could instantly see what inventory was sitting in dark retail locations and turn those stores into shipping centers. When guests shopped online, we could promise them products we actually had, not phantom inventory that existed only in outdated spreadsheets.

The technology let us do things our competitors couldn't. Ship from the closest store instead of a distant warehouse. Show real inventory online without disappointing guests with out-of-stock surprises. Complete inventory counts in hours instead of weeks when stores reopened.

Same story with our cloud infrastructure. We'd started that migration years before anyone heard of COVID-19, not because we predicted a pandemic, but because we believed in owning our ability to scale. When digital traffic exploded and everyone else was scrambling for server capacity, we just turned up the dial. During the peak of lockdown shopping, the week of Black Friday, we processed nearly half a billion digital orders. Our systems handled it like it was Tuesday. We were fully in the cloud just after the pandemic ended, just about five years after we began.

Those weren't lucky breaks. They were the compound returns of strategic patience. We'd chosen to invest in capabilities before we needed them, to build foundations that could support futures we couldn't fully envision.

The pandemic tested everything we'd built—the technology, the teams, the culture of ownership and truth-telling. What held wasn't just the code or the servers. It was the belief that we could figure it out together, that problems were puzzles to solve rather than disasters to survive.

Positivity Matters

Holiday 2020, from just before Thanksgiving to just after Christmas, was unlike any before. There was no roadbook for peak performance in the middle of a global pandemic. Demand patterns shifted weekly, carrier networks were overloaded, and inventory constraints loomed large. Guest expectations, however, remained high, if not higher, than ever.

Sitting in that first war room meeting, looking at the impossible task ahead, I thought about Coach Charlie.

But what made it work wasn't the process. It was the culture that enabled the process.

Charlie led my seventh-grade girls' basketball team. We didn't win a single game that season. Not one. But Charlie never gave up on us. He showed up with a southern drawl, a whistle around his neck, and a bag full of sayings that could make us laugh whether we wanted to or not. "Ladies, there ain't no 'I' in team!" he'd call out. Or when someone went down dramatically, "Sacrifice your body for the team!"

The scoreboard never tipped our way, but we felt seen. Encouraged. Trusted.

Years later, facing Holiday 2020 with my team exhausted from nine months of crisis, I finally understood what Charlie had been teaching us. Positivity matters. Presence matters. Team matters. Even when the results aren't there yet, people remember how you showed up.

You can't lead transformation if you only show up when things are going well. You especially can't lead transformation when you abandon people during the hardest season they've ever faced.

So we pulled together a cross-functional coalition that operated like a mission control center. Our approach to peak became militaristic in its rigor and agile in its execution.

Culture as Advantage in Pandemic

Daily war rooms became the heartbeat of our operations. Tech, Guest Services, Omni-Channel, Supply Chain, and

Retail Ops leaders met every morning to review real-time metrics—order volumes, fulfillment times, system uptime, guest escalations.

These weren't status meetings. They were problem-solving sessions where people surfaced issues early, asked for help without losing face, made decisions without waiting for approval. The truth-telling culture we'd built meant problems got addressed before they became crises.

When someone hit their limit, the team stepped in. No one was expected to be a hero at the expense of their well-being.

Teams showed up with grit, grace, and unwavering commitment. We didn't just hit our targets. We rewrote the playbook on managing peak in a hybrid, uncertain, high-stakes environment.

We practiced peak purchasing time like it was a sport—and every person played to win.

Culture is the ultimate competitive advantage. Our competitors had the same technologies, the same cloud platforms, the same fulfillment vendors. But they didn't have the culture of trust, ownership, and collaboration that let us move at the speed of change.

When Values Become Survival Tools

Perhaps the most remarkable part of our COVID story happened inside the team. In the chaos of shifting priorities, urgent launches, and impossible timelines, our culture held.

When people shifted to remote work, they brought their values home with them. Courage looked like engineers speaking up about burnout and boundaries. Connection showed up in virtual yoga classes, daily stand-ups with check-ins before tasks, teammates covering for each other during crises.

The vulnerability principle that had been so hard for me to learn personally became organizationally essential. Leaders who admitted they didn't have all the answers created space for teams to innovate. Managers who acknowledged their own fear gave permission for others to be human too.

Engineers shipped code from kitchen tables. Product managers ran standups from living rooms with toddlers in the background. Leaders navigated personal loss, anxiety, and burnout while still showing up for their people.

One of our senior directors hosted a weekly "human hour." No agenda, just presence. People shared stories of caregiving, fear, joy. We listened. Laughed. Sometimes cried. It wasn't about productivity. It was about belonging.

We had no roadmap. But we had trust.

We also had to innovate how we connected. We built digital platforms where guests could work out together, shop with personal stylists, connect with each other— all from home. We rebuilt chat support infrastructure to offer round-the-clock assistance—automated for routine questions, human agents for moments needing empathy. We added AI-powered search that learned what guests wanted. Not to replace human judgment, but to give our team better information for conversations that mattered.

Guest feedback told us it didn't matter where they interacted with us as long as they felt seen.

That winter, I saw engineers cry, not from burnout, but from pride. They had built something that mattered.

Leadership Alignment

We had a vision larger than sales—we wanted to keep our store teams paid even though stores were closed. This was our front line and we truly wanted to take care of those who take care of our guests. To make that possible, we set aggressive expense management targets across the company.

Our best vendor partners didn't skip a beat, dropping their rates and continuing to work. The executive team took a 20 percent pay cut. The board of directors gave up their cash retainer.

These weren't symbolic gestures. They were choices that made the difference between keeping people whole or letting them fall. Between living our values or just talking about them. We didn't lay off a single employee during the pandemic. Not a single employee, even though all our stores were closed.

Culture isn't just what you say when things are good. It's what you do when everything's on fire and the easiest path is to protect yourself first.

Strategic Shifts That Endured

The legacy of our COVID response is not in the quick wins. It's in what we carried forward.

Global resiliency planning was embedded into our operating rhythm. What began as crisis response now informs how we prepare for peak seasons, supply chain disruptions, and regional instability.

Technology culture became an advantage. The practices that helped us survive the pandemic—cross-functional collaboration, open feedback loops, human-first leadership—are how we worked every day. We kept what made us strong.

These weren't temporary adaptations. They were long-term investments in a more agile, guest-focused, and resilient future. And they were all enabled by the culture we had built before the crisis hit.

None of this was preordained. It was built. One pivot, one conversation, one brave decision at a time. By people who trusted each other enough to move fast, fail smart, and keep pushing forward.

When the world paused, we accelerated. lululemon's COVID response wasn't just about holding the line. It was about showing what was possible when a values-led company trusts its people and its technology.

We didn't just serve guests. We redefined how. We didn't just survive. We evolved. And that acceleration didn't stop when the lockdowns lifted. It became our new normal.

The pandemic proved that the authentic foundations we had built—the culture of trust, the integrated technology systems, the focus on people first—weren't just nice-to-have values. They were competitive advantages that enabled us to not just weather the storm, but to emerge stronger.

When crisis comes, technology can break. Processes can fail. Markets can disappear. But culture endures. And if you've built it right, it becomes the engine that drives you through uncertainty and into whatever comes next.

That's the real lesson of 2020. *Culture isn't a soft skill.* It's the hardest skill of all. And in the end, it's the only thing that scales.

The pandemic proved our culture could withstand crisis and adapt at speed. But success can create dangerous assumptions. Remote work had succeeded so well that I began to believe proximity was optional, not essential. I opened remote hiring across time zones, bringing in incredible talent from anywhere. What I didn't yet see was how that decision would test the very foundation we'd worked so hard to build. That lesson would come later, and it would cost us.

The Truth About Change

The pandemic proved our culture could withstand crisis and adapt at speed. But success can create dangerous assumptions. Remote work had succeeded so well that I began to believe proximity was optional, not essential. We opened remote hiring across time zones, bringing in incredible talent from across the globe. It seemed like a pure win.

Until we came back to the office.

We didn't come back because we distrusted remote work. People had proven they could be productive from anywhere. We came back because our culture, the one we'd spent years building, was designed around spontaneous collaboration, walking conversations, reading energy in rooms. We could have redesigned for remote, but that would have meant rebuilding our culture from scratch.

Some companies went fully remote and made it work brilliantly. The difference wasn't remote versus in-office. It was whether leaders made an intentional choice based on their actual culture and work, or whether they defaulted to what felt comfortable or what others were doing. We chose hybrid because our culture was built on collaboration that happened best with some face-to-face time. But that was our answer, not the answer.

But when I asked people to return to the office, I discovered the complexity I hadn't anticipated.

Some people welcomed it. They'd been isolated, craving the human connection, the energy of working side-by-side to solve problems. They missed the spontaneous conversations, the ability to read a room, the feeling of being part of something together. They had felt isolated. For them, coming back felt like coming home.

But others had established new lives that worked for them—lives built around family rhythms, healthier routines, and hard-won balance. Coming back didn't just mean restarting a commute; it meant dismantling those lives. New remote hires felt excluded from culture being rebuilt in rooms they couldn't enter. Some employees felt penalized by traffic, costs, and lost flexibility. Managers were caught in the middle, torn between policies and the people they cared for.

The result wasn't universal resistance, but instead it was division. Some people thrived being back. Others complied but resented it. And that division itself became the problem. What we got wasn't rebellion, but a fracture. We got compliance without commitment from some, enthusiasm without understanding from others. The culture that had been our engine started to sputter. The impact showed up everywhere. Information flowed unevenly. Remote team members began to hesitate before speaking up, unsure if their voices carried the same weight. In-office teams grew frustrated at carrying more of the culture load. Leaders sensed the strain but struggled to repair it. Trust, the very thing that had carried us through crisis, began to erode.

I had learned this lesson before. I had closed San Francisco and Portland because scattered teams couldn't build the kind of trust and collaboration we needed. I had consolidated in Seattle and Vancouver because proximity creates connection, and connection enables speed. But in the excitement of pandemic innovation, I convinced myself that culture could transcend geography without intentionally redesigning for it.

I was wrong. Not because remote work doesn't work. It does for many companies. But I hadn't been intentional about what our specific culture needed.

The problem wasn't that we asked people to come back. The problem was how we talked about it.

We sold the upside. We were scripted with consistent talking points. We talked about collaboration, innovation, spontaneous hallway conversations, the magic of in-person connection. All of that was true. But we didn't talk about the other side. We didn't acknowledge that people were losing the flexibility they'd built their lives around, the family time they'd reclaimed, the commute costs they'd eliminated, the routines that had made them healthier and happier.

When you only tell people half the truth, they don't feel led. They feel managed. They feel spun.

People can handle hard truths. What they can't handle is feeling like you're pretending there's no cost to what you're asking them to do. When things happen TO them without being brought on the journey, when you skip over the sacrifices and only highlight the benefits, trust breaks down. Not because the decision is wrong, but because the communication is incomplete.

This is why compliance without commitment happens. Not because people are resistant to change, but because they don't trust leaders who won't acknowledge the full picture.

This matters now because we're making the exact same mistake with AI.

Leaders are breathlessly selling the upside of AI. Efficiency! Innovation! Transformation! Augmentation, not replacement! The future is here and it's amazing!

But they're not talking about what people are truly experiencing. The fear that their skills are becoming obsolete. The anxiety about job security. The loss of work that gave them meaning and identity. The feeling that they're being automated away while being told to be excited about it.

When you pretend there's no downside, when you skip over the real human cost of technological disruption, you lose trust before you even start. People aren't stupid. They can see what's happening. And when you refuse to name it, they assume you're either naive or dishonest.

Neither builds trust.

The lesson from return to office isn't about remote versus in-person work. It's about what happens when leaders ask people to embrace change without acknowledging what that change actually costs. It's about the erosion of trust that happens when you treat people like they need to be sold rather than leveled with.

AI is the biggest workplace transformation most of us will see in our lifetimes. It will create incredible opportunities. It will also displace jobs, eliminate roles, and fundamentally change what work means. Both things are true. And if we only talk about one side, we'll lose the trust we need to navigate the other.

People don't need you to sugarcoat change. They need you to be honest about what it costs. They need you to bring them on the journey, not just announce the destination. They need to know you see the sacrifice, not just the benefit.

That's what I got wrong with return to office. And that's what too many leaders are getting wrong with AI right now.

The question isn't whether to pursue transformation. It's whether you're willing to tell the truth about what it takes.

For Your Leadership

The pandemic proved that culture isn't what you fall back on when everything breaks. It's what makes rapid transformation possible in the first place. AI transformation will test your culture the same way.

Architecture Decisions Create Options or Constraints: Years of platform investment paid off when we pivoted omnichannel overnight during the pandemic. The decisions you make today about data infrastructure, model governance, and technical foundation will either enable or limit you when pressure hits. Every architecture choice is either building future capability or creating future debt.

Trust Enables Speed: We moved fast during the pandemic because we'd distributed decision-making authority years earlier. Teams didn't wait for permission. Instead, they acted. Distribute authority before you need speed.

Resistance Reveals What People Value: Return-to-office pushback wasn't about offices. it was about autonomy and trust. Listen to resistance as data about fears you haven't addressed, not ignorance you need to overcome.

Culture Carries You When Plans Fail: When disruption moved faster than our plans, culture carried us forward. The culture you build now determines whether you adapt or break when transformation gets hard.

When crisis comes, technology can break. Processes can fail. Markets can disappear. But culture endures. And if you've built it right, it becomes the engine that drives you through uncertainty and into whatever comes next.

The real lesson of 2020 is that *culture isn't a soft skill.* It's the hardest skill of all. And in the end, it's the only thing that scales. The pandemic proved our culture could withstand crisis. But the return-to-office experience taught me that culture that works in one context doesn't automatically translate to another.

The foundation we built was real—real enough to carry us through the chaos of 2020. But it wasn't infinitely flexible. Now I needed to test whether culture could expand across continents while staying authentic?

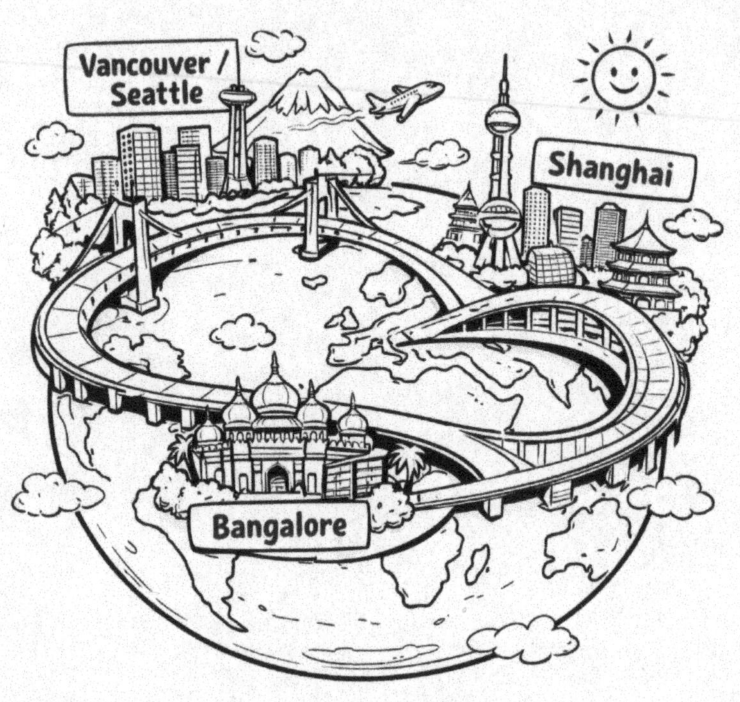

Vancouver /
Seattle
Shanghai
Bangalore

CHAPTER 10

Building Bangalore

We were walking outside, maintaining the six-foot distance that had become second nature by early 2021. Seshan and I had been venting about vendor frustrations—the kind that had multiplied during the pandemic. Missed deadlines while we were racing to keep the business running. Communication gaps when we needed instant collaboration. Quality issues at exactly the wrong moments.

The pandemic had exposed just how fragile our vendor dependencies really were. When everything went remote overnight, when stores closed and digital became not only our main channel but our only channel, when every hour mattered, we couldn't afford to wait for contractors who worked only during business hours halfway around the world. We also realized we were operating with one hand tied behind our back. Our technology organization was essentially confined to a single time zone. When our Vancouver and Seattle teams clocked out at 5 PM, work stopped. We had no follow-the-sun capability, no way to keep development and innovation running around the clock to match our global business that never slept.

"I wish we had our own hub in India," I said, almost without thinking.

Seshan stopped walking. Even with masks and distance, I could see the shift in his expression.

"We can do that."

I turned to him, eyebrows raised. In that moment, with the world still locked down and uncertainty everywhere, we began imagining what was possible.

That casual comment during a pandemic walk would spark one of the most transformational initiatives of my tenure at lululemon. But more than that, it would become a masterclass in how inclusion isn't just the right thing to do, but rather the smartest business strategy you can pursue.

But more than that, it would become a masterclass in how inclusion isn't just the right thing to do, it's the smartest business strategy you can pursue. India had just overtaken the United States to become the world's largest base of open source contributors on GitHub. We weren't just checking a diversity box. We were tapping into the deepest pool of technical talent on the planet, in a country that had evolved from the world's back office to the world's innovation engine. India produces more engineering graduates than any other nation. It's home to the largest pool of data scientists, AI researchers, and cloud architects outside of Silicon Valley. By 2030, one in every three new developers globally will come from India. Companies that treated India as just a place to save money were missing the point entirely. This wasn't about cost. It was about capability. India wasn't a secondary option. It was a strategic competitive advantage.

The Breaking Point

In late 2020 I was looking at spreadsheets that told a story I didn't like. We had doubled our employees since 2018, but nearly three-fourths of our labor budget was going to contractors. Hundreds of critical positions were held by people who didn't have a long-term commitment to the company.

The breaking point came when one contractor managing a critical project demanded to be hired as an employee—with a title that didn't match his actual ownership and a salary outside our established range. The request came to me as an exception because of how critical his work was.

I was looking at two bad choices. Either we hurt business outcomes by losing him, or we set a precedent that told people we would abandon our workforce principles if a person had leverage. I didn't like either option, so moved a different resource into that role and exited the contractor.

Business continued, but not without disruption. The new resource took three weeks to get up to speed, and we missed one sprint deadline. Some leaders questioned whether I'd made the right call at the wrong time. The contractor was well loved and had done great work. But what was reinforced was that we had contractors with too much leverage, and that giving in would only make the problem worse.

I'd watch the same cycle repeat. First, hire contractors to move fast on a project, then lose them when the contract ended, and then hire new contractors to rebuild what we'd just lost. It was expensive and exhausting, but worse than that, it was strategically dangerous.

Meanwhile, our business was global and operated 24/7, but we were leveraging talent from essentially one time zone. When our Vancouver and Seattle teams went home, work stopped. When issues arose in Asia-Pacific overnight, they waited until morning to get addressed.

But the deeper problem was that we were outsourcing our future. The company had a voracious appetite for more technology. We needed personalization engines, demand forecasting, inventory optimization, dynamic pricing. These were no longer nice-to-have features. They were table stakes. Yet we couldn't hire the right talent fast enough in North America.

From Conversation to Conviction

When Seshan said "we can do that," something clicked. This wasn't just about reducing contractor costs or spreading work across time zones. India had become the world's largest producer of top technology talent, and most Western companies only accessed that expertise through vendor relationships. We had the opportunity to build these capabilities as core competencies.

What started as a frustrated comment became an eight-week sprint to prepare for the board of directors. I knew we didn't have the luxury of a multi-year planning process. Our 2021 strategic initiatives were already assuming we'd have this increased capacity—to the tune of nearly $100 million in additional capital investment.

We used a partner to help us with market entry. That contract was signed in February 2021. By June, Seshan had moved to Bangalore and we had our official India Tech Hub grand opening. From the beginning Seshan and I knew we wanted to build something special, something with a culture that aligned with the company's goals of reflecting the diversity of the communities we serve and operate in around the world by 2025[4]. To get there, we needed authentic inclusion.

Learning India from the Ground Up

Before we could build authentic inclusion in the India Tech Hub, I needed to understand what authentic partnership looked like. Years earlier, as I began leading technology transformations, I found myself working with vendor partners in India. For years, I experienced India through the lens of a Westerner doing business there. My vendor partners would prepare special events for me and my team, carefully curated experiences designed to be "safe for Westerners." We'd visit archaeological sites where guides spoke perfect English and shared sanitized histories. We'd see elephants at wildlife sanctuaries with paved paths and gift shops. When I mentioned wanting more adventure, they'd arrange slightly more elaborate versions of the same filtered, or carefully managed, experiences.

I knew from Ethiopia what unfiltered meant, and I knew I was getting a filtered view. The India I was seeing felt like a theme park version of itself—beautiful, but somehow hollow. The relationships felt transactional, polite, but distant. Over time, I realized I needed something more authentic if I was going to truly understand the people I was working with.

It wasn't until I let my team actually guide me that I started to see the real India. The shift began when I stopped planning the itineraries and started saying yes to their suggestions. Venki and Seshan took me to their ancestors' temple, not a tourist destination, but a living place of worship where their families had prayed for generations. The stone floors were worn smooth by centuries of bare feet, and the air was thick with incense and devotion. They explained the significance of each carving, each ritual, with a reverence that made me understand this wasn't just culture. It was identity.

Seshan, myself, Venki in Coorg, India

Seshan invited me into his parents' home for dinner, which felt more welcoming than any five-star hotel. His mother, who spoke no English, communicated through smiles and gestures as she served us homemade *dal* and *chapati*. His father was warm and so welcoming. I felt at home.

But the real learning began when I started exploring beyond the business centers and tech campuses.

Walking the Streets, Learning the Stories

My local team could see I was genuinely open to learning beyond the business centers, so they began suggesting we explore their neighborhoods. We walked through residential areas where families gathered on front steps in the evening, children played cricket in narrow alleys, and vendors sold fresh fruit from bicycle carts. I tried street food despite security team protests—samosas from roadside stalls, real chai tea, spicy chaat that made my eyes water but taught me about layers of flavor I'd never experienced.

The conversations were revelatory. I spoke with high school students about their dreams, their view of technology careers, what they thought about working for international companies. Their perspectives challenged my assumptions about ambition, family obligation, and success. Many were brilliant but felt pressure to choose between pursuing their own interests and supporting extended families. Others spoke about the gap between what they learned in school and what employers actually needed.

Visiting a school outside Bangalore

I visited local villages outside Bangalore, places where our employees' families lived. I met grandmothers who had never used technology but whose granddaughters were building it. I saw the tension between tradition and rapid change, the way families navigated modernization while preserving what mattered most to them.

The Hidden Talent Pool

Through Seshan's network, I connected with women's rights activists, including a friend who had spent decades fighting for educational opportunities for girls in rural areas. She could recall the battles while keeping daughters at home, and how that was slowly changing as technology created new opportunities. Her insights helped me understand that we had stumbled onto a competitive advantage hiding in plain sight.

In India's overall workforce, women represent just 24% of workers. In the IT sector specifically, that number rises to 34%. But here's the magic part: nearly 47% of IT and computing postgraduates in India are women. Think about that gap. Almost half the graduates, but only a third of the workforce. That's not a pipeline problem. That's millions of educated, capable women who either can't find companies willing to hire them or can't find environments where they can succeed.

While most companies were complaining about talent shortages, we were looking at the largest untapped talent pool in the world's second-largest developer ecosystem. Women weren't just available. They were educated, eager, and systematically overlooked by companies stuck in traditional hiring patterns. We weren't just building a tech hub. We were accessing talent our competitors didn't even see.

Building for What Others Built Out

Years before we opened the India Tech Hub, an Indian vendor partner asked me to speak to their women's group. I didn't just speak. I listened. What I learned changed everything. I thought I brought. I knew about barriers to women in technology.

The women told me about pressures I'd never considered. Many faced expectations to marry young and prioritize family over careers. Even highly educated women often stepped back after marriage or childbirth, not by choice, but because of family pressure. In joint family systems, they were expected to care for not only children but also in-laws, making the demanding schedule of tech careers nearly impossible.

One woman told me she'd been promoted to a leadership role, but her mother-in-law said it reflected poorly on the family that she worked such long hours. "A good wife should be home by evening," she was told. Another described how her brilliant ideas in meetings were dismissed, but when a male colleague repeated the same suggestion, it was praised.

I learned that India has one of the highest female workforce exit rates after maternity, not because women lose interest in their careers, but because the infrastructure and cultural support aren't there, with inadequate childcare infrastructure and cultural expectations cited as primary barriers.

And the barriers go deeper. Research from India's National Human Rights Commission revealed that 96% of transgender individuals are denied jobs and forced into low-paying or undignified work like begging or sex work. Only 6% of transgender people were employed in the formal sector[5]. Around half the transgender population never attended school, and those who did faced severe discrimination.

Now I understood what we were up against. The talent pool was massive. The capability was undeniable. But accessing that talent meant building something fundamentally different. We couldn't just hire women and LGBTQ+ individuals into the same old system and expect different results. We had to create an environment where they could actually succeed.

This wasn't just about numbers. It was about infrastructure. In India, working mothers have virtually no support systems. The "vision of gender balance" opens doors, but performance keeps them open. What we needed wasn't speeches about diversity. We needed childcare support, flexible work arrangements, parental leave policies that actually worked, leadership training, and a culture that valued results over face time. We needed to design the tech hub from the ground up with these realities in mind.

One person told me about the exhausting mental load of constantly monitoring their speech and behavior, never knowing when a slip might reveal their truth. I knew that fear well and had been able to walk away from it. Another described the isolation of not being able to bring their partner to company events or even mention them in casual conversation. Again, relatable.

But what struck me most was the tension. These women and LGBTQ+ individuals were more educated and ambitious than ever, but systemic barriers meant many couldn't fully realize their potential. They weren't asking for charity or special treatment. They were asking for the chance to contribute their full capabilities without having to hide who they were.

That conversation became my blueprint. Our goal wasn't just building a tech hub. It was proving that a fully gender-balanced workforce and inclusive benefits weren't just good values. They were good business strategy. We needed to create an environment so compelling that talented engineers would choose us over companies that couldn't see or didn't value what these people actually needed to succeed.

The numbers told part of the story. But the real work was building systems that let talent thrive. Creating an environment where the 47% of women could actually build the careers they'd trained for. Where performance mattered more than conformity. Where people could bring their whole selves to work without fear.

This work isn't meeting potential. It's unleashing it. Sometimes people need encouragement. Sometimes they need skills training. But most often, they just need someone to remove the barriers that shouldn't have been there in the first place.

The Leader Who Taught Me to See

Before I could build inclusion at scale, I needed to understand what it meant to truly see people. That lesson came from a bench in a backyard, years after I thought my education was complete.

A few years after Ermias came home, some MBA alumni organized a reunion. I was running late and texted that I might not make it. The host replied, "Ali is still here and will wait for you."

I hit the gas. Dr. Ali Tarhouni had been my economics professor at the University of Washington. He later went on to be the acting interim Prime Minister in Libya. I hadn't thought he would remember me. I was one student in one cohort years ago. But by the time I arrived, the party was dissolving, and Ali was still there. We found a bench in the yard.

He asked me how I was. I gave a surface answer. He looked me in the eyes and asked again.

This time I told him the truth. Cindy and I had just adopted Ermias from Ethiopia. I was carrying the weight of his story, the loss he had endured, the reality of taking him from a family that loved him but could not care for him.

Ali listened. Then he told me about Libya. About leaving the comfort of his family in Seattle to return to a country he loved that was tearing itself apart. He'd been exiled for opposing Gaddafi, sentenced to death in absentia. During the revolution, he returned to serve as Minister of Finance and Oil, guiding the country through its fragile first days. He understood loss and love and impossible choices in ways most people never will.

We sat together in that understanding, two people who had chosen to embrace complexity because love demanded it.

Then he asked, "You went to Seattle Pacific University, didn't you?"

I was stunned. How could he possibly know that?

"I was the professor sitting silently in the back of the room during your admissions simulation," he said. "I hand-picked each person for the program. I know all of my students."

Years earlier, I'd scrambled to apply to UW's first evening MBA program with just two days' notice. The admissions day felt like a gauntlet—group exercises, timed essays, high-pressure simulations. I left convinced I'd blown it. But I'd been accepted, and the program changed my life.

Sitting on that bench, understanding that Ali had chosen me, the lesson crystallized. It wasn't just that he'd seen potential in me. It was that he cared enough to truly see me in the first place.

That became my blueprint for India. Not hiring for credentials alone, but seeing the person underneath. Not waiting for people to prove themselves, but believing they were extraordinary before they'd shown you. Not leading from a platform, but leading from presence—being close enough to see what others miss.

When I insisted on walking the streets of Bangalore instead of staying in conference rooms, I was following Ali's example. When I made knowing each person's story non-negotiable, I was applying what he'd taught me, which was leadership isn't about having all the answers. It's about *seeing people* clearly enough to help them find their own.

The Privilege of Bold Ambition

I also recognize the privilege of working in a company that set bold ambitions beyond growth, but also ambitions of impact. lululemon's 2020 Impact Agenda stated, "By 2025, the lululemon workforce will reflect the diversity of the communities we serve and operate in around the world." It wasn't about quotas or meeting numbers but about understanding where talent was being overlooked and removing barriers for that talent to achieve their best. This resonated to my soul.

That commitment gave me the platform and permission to ask harder questions about inclusion and to design systems that matched our values. It reminded me that diversity alone doesn't create excellence, but access and accountability do. We built teams where performance and well-being weren't at odds, where equity meant everyone had a fair shot to succeed and was equally responsible for delivering results.

That commitment gave me the platform and permission to ask harder questions about inclusion and to design systems that matched our values. It reminded me that diversity alone doesn't create excellence, but access and accountability do. We built teams where performance and well-being weren't at odds, where equity meant everyone had a fair shot to succeed and was equally responsible for delivering results.

But understanding lululemon's commitments was only half the equation. I still needed to understand India on its own terms. That education continued in unexpected places.

The Pickleball Court Revelation

One afternoon, knowing how much I love pickleball, Seshan suggested we visit courts an hour outside the city. As we drove through increasingly unfamiliar neighborhoods, past street vendors and children playing cricket, he warned me, "this isn't the India that Westerners typically see. These are working people, not tourist guides." Sounded perfect for me.

There were three courts—simple concrete slabs surrounded by chain link fence. But the moment we arrived, I was welcomed based solely on my willingness and ability to play. They primarily spoke Hindi, but sport created its own language. We played hard and celebrated good shots. When I left, one of their best players, a woman, shook my hand and said "backhand"—the universal need for improvement that transcends culture.

I kept going to the morning practices even when Seshan couldn't. One hot Bangalore day I knew I hadn't drunk enough water to be playing, but chose to keep playing anyway. My dad's voice in my head saying "there's no crying in baseball" had apparently translated to "there's no quitting in pickleball"—even when quitting would have been the smart call.

In the middle of a game, I felt my stomach turn. I kept playing. "Just a few more points, hold it together Averill," I told myself. Until I started to collapse. I ran to the side of the court and threw up—everything! Play stopped immediately on all three courts. I was horrified. Everyone from each court ran to me, even as I was begging them to please keep playing. One person ran home and brought back salt water for me to drink. Another held my hair and rubbed my back. Everyone showed such genuine concern. I was mortified, of course. I hated being the frail white person who just suffered heat stroke. But, at the same time, I felt care from strangers in a way that was foreign to me. I saw their faces, they were worried and I became more important to the moment than their play. When I recovered, they told me their secrets: hydrating the day before, cold towels around their necks, dark hats to deflect the sun.

Six months later, Seshan and I partnered together in a tournament in Goa. I was proud to say that he and my female partner, who was also Indian, both got heat stroke but I did not. I'd learned. Mortification is an excellent teacher.

These experiences taught me something crucial about building global teams. Competence and character translate across cultures. Respect is earned through showing up authentically, not through titles or corporate positioning. It's about learning and growing versus teaching. Humility. Sometimes people put egos together for purpose, and nothing else matters. The foundation for inclusion isn't policy—it's creating conditions where hierarchy, background, and personal identifiers fall away, and purpose becomes enough. Something people come together for purpose, not a place where diversity was celebrated, but a place where genuine human connection was center.

Every country I travel to now, I pack my paddle. Not just for work, everywhere. Pickleball is how I make real connections. The court strips away hierarchy and language barriers. You learn more about someone in ten minutes of play than an hour of small talk. And you can laugh.

From Understanding to Strategy

North American leaders weren't just responsible for their local teams, they were responsible for the entire global team's success. You couldn't mentally separate "my team" from "the India team" because it was all one team.

When we built "one team, one culture," it was because I had experienced how authentic relationships could transcend geography and difference. I had seen how trust could be built through presence, curiosity, and genuine care for people's full humanity.

Inclusion as Strategy

From the beginning, we knew this couldn't be just another offshore operation. We needed to think fundamentally differently about how we built the team.

Seshan and I were committed to understanding why talented women were leaving tech, removed those barriers, and built award-winning teams. We learned while women accounted for about 43 percent of technology graduates in India[6]—the highest rate globally—yet only 14 percent of STEM jobs in the country were held by women[7]. India has one of the world's lowest female workforce participation rates, with women's employment dropping sharply after marriage[8] while men's careers accelerate[9].

We focused on removing barriers through flexible work arrangements, support for career breaks, inclusive benefits, and leadership development programs.

The results spoke for themselves—by 2023, **women represented 46 percent of our workforce**. Three years in a row, WEQUITY recognized lululemon as one of the "Most Inclusive Organizations for Women in Tech." But the real was watching talented people step into roles they might never have had elsewhere, building networks of support, and proving that when you remove barriers, innovation follows.

We also built our benefits package around inclusion in ways that went beyond industry standards. Support for IVF treatments, gender transition procedures, partner benefits for LGBTQ+ employees—even though same-sex marriage wasn't legally recognized in India. We weren't just following local customs; we were taking a leadership position because we believed talent thrives when people can bring their whole selves to work.

The real cultural foundation was our "one team, one culture" philosophy. This wasn't about teaching India what lululemon was about. It was about learning from India and building something together. We would operate as a single, integrated team, not as headquarters and satellite office.

That philosophy was tested immediately. Would we really treat Bangalore as equal partners, or would it just be nice words covering up the same old power dynamics? The early decisions we made—about meeting times, language, authority, and respect—would determine whether we built authentic inclusion or just better-marketed outsourcing.

Out With Pride: Setting the Cultural Foundation

Before the official inauguration of the India Tech Hub, before we'd even fully opened our doors, we hosted an "Out With Pride" event. Over 200 people gathered—employees, community members, the public—for yoga, tai chi, smoothies, and celebration. This wasn't a corporate diversity initiative. It was a statement about who we were going to be from day one.

I stepped onto the stage in a country where homosexuality had been decriminalized just five years earlier. Standing there, I talked about being gay. About what it had cost me to hide for so many years. About why creating a workplace where people could bring their whole selves mattered more than any business metric.

The risk was real. Not everyone in the audience agreed with my life or my choices. But I wasn't there to get everyone's approval. I was there to send a message to every person we were about to hire that *you don't need to hide here*. You don't need to perform a version of yourself you think we want to see. Bring who you are. That's who we hired.

The photos from that day show something I'll never forget—the faces of people who'd never seen a senior executive be that honest, that vulnerable, that public about their identity. Several women approached me afterward and said they were reconsidering whether they could work for us. Not because I was gay, but because if I could be that open, maybe they could too.

That event became the foundation everything else was built on. The 46% women workforce. The inclusive benefits. The psychological safety that let people surface problems early. None of it would have worked if we'd started with policy instead of humanity.

The Saree Moment—and What It Really Represented

When it came time for the official inauguration, Seshan approached me with an invitation that made me immediately uncomfortable.

"Julie, the team is asking if you would wear a saree for the opening ceremony."

My first instinct was to decline. Politely, but firmly. I didn't want to appropriate their culture. I didn't want to look like a Western executive playing dress-up, performing respect rather than earning it. I'd spent years watching leaders do "cultural appreciation" photo ops that felt more like tourism than genuine engagement.

But more than that, I was afraid of being disrespectful. As a white person coming into their space, their culture, I didn't want to get it wrong. I didn't want to be another outsider who thought putting on traditional clothing made them an insider.

"I don't know, Seshan. Won't that seem like I'm trying too hard?"

He smiled. "The team has been talking about it for weeks. They want to share this with you. It's not about you trying to be Indian. it's about sharing the culture."

I thought about it overnight. Asked Cindy what she thought. "You love India. You love their culture, their streets, their food, their humanity. This isn't performance. This is participation."

The distinction landed. Performance is about how you look to others. Participation is about showing up for what matters to them.

A woman named Shilpa, someone who became a trusted advisor over the years, took me shopping. Fabric after fabric laid out on the table. Not exactly the try-on experience I was used to. Eventually we found it. Blue and gold with a gold shirt underneath. The morning of the inauguration, she helped me get ready. She wrapped it around me, tucking and folding with practiced hands, laughing at my questions about how anything stayed in place without buttons or zippers.

Four safety pins. That's all that held 6 or 7 yards of fabric onto my body. I was doubtful. But she assured me.

When I walked into the room for the ceremony, the energy shifted. I felt self-conscious for about three seconds—then I saw their faces. Joy. Surprise. Happiness.

My team looked at me like I'd given them something they didn't know they needed.

One brave woman was the first to approach. "Can I take a picture with you?" Then her friend joined for a shot of the three of us. Then their whole team wanted in. Then each person individually. Then pairs of friends. Then an entire other team lined up. It kept going—person after person wanting to capture this moment, wanting to share it with someone who mattered to them.

Standing there as the line grew, I felt welcomed. They appreciated the respect I had shown, which was exactly what I'd been hoping for. Later, I learned many of them sent those photos to their parents—showing them that a North American CIO was with them, as they were and as I was.

It was a small gesture that sent a powerful message. We were there to learn and build together.

Celebrating in India with Women in Tech

Rewiring the Language of Inclusion

Within months of opening, I made a discovery that would prove crucial—and I made it the hard way.

On my first trip to India after we launched the Tech Hub, I was working with North American teams from Bangalore. That's when I felt it. *I was "off-shore."* It was assumed that I would flex to meet "real business hours"—their hours. Never mind that I was the CIO sitting in our newest office with our newest team. The default was North America, and I was expected to adjust.

Sitting in our offices in India and listening to how people talked, I kept hearing the same problematic terminology of "on-shore" and "off-shore." North America was "on" and India was "off," immediately creating a hierarchy where one location was the default and the other was secondary. When I was referred to as "off-shore," the realization struck me. I was in India and I was the one who was off, the one who didn't fit in. It was also interesting to me that when I was in China, Hong Kong or Europe, the word "off-shore" was never even considered.

If they were doing this to me—the CIO—what were they doing to everyone else on the India team?

I talked with the global team about the importance of language and it's power—the unintended messages that were sent with just one word. "Offshore" had come to mean distanced and lesser than. I talked about this concept openly with small groups and the larger team, inviting everyone to understand the subtle word choices that imply privilege and choice. I shared from my own perspective as I had used that language for years with service providers where it is industry standard. But experiencing it myself changed everything. The team quickly stopped using that term, and instead simply said "our ITH (short for India Tech Hub) team."

But language was just the start. I noticed that the burden of time zone differences always fell to the India team. They were the ones staying late or getting up early to accommodate North American schedules. This couldn't continue if we truly believed in "one team."

I challenged North American leaders to answer "how are you solving for this burden?'

No one pushed back directly. Instead, they chose silence. They nodded in meetings, agreed in principle, but I couldn't tell if anything was actually changing from 8,000 miles away.

It wasn't until a trip to India with all the managers and their managers in one room where I heard the reality that some teams were flexing, some weren't. The India team members were candid when we were all together in person. Some of their North American counterparts were splitting meeting times fairly. Others weren't budging.

I decided to amplify what was working rather than shame what wasn't. I asked one of the managers who was getting it right to share his approach with the group.

"Every week we switch times for our stand-ups," he explained. "One week it's off-hours for North America, the next week it's off-hours for India. Nobody's comfortable all the time, but everyone's uncomfortable equally."

The room got quiet. It was such a simple solution, but half the teams hadn't thought to do it. Or hadn't wanted to.

After that meeting, the message was clear that we all owned this challenge together. I restructured accountability to match our values. North American leaders weren't just responsible for their local teams, they were responsible for the entire global team's success. You couldn't mentally separate "my team" from "the India team" because it was all one team.

Breaking Norms

But our inclusion efforts didn't stop with females. We recognized that true inclusion meant addressing barriers that affected everyone, including those that traditional corporate environments rarely acknowledged.

In India, there's a Hindi proverb "मर्द को दर्द नहीं होता", or *Mard ko dard nahi hota*, which means "a man does not feel pain," made popular in Bollywood films over the years. It reflects a stereotype that men are expected to not show weakness or express pain. They are expected to be stoic providers who never cry or admit struggles without being seen as lesser than those who do not.

We decided to challenge this directly.

We launched "Men Ki Baat," which translates to "Men's Talk" in Hindi, where men could talk about mental health without judgment. No hierarchy, no performance, just space to be human. We brought in psychologists and doctors to address issues that men typically suffer through in silence. Issues such as workplace pressure, the expectation of always being strong, family expectations, the burden of being the sole provider.

The first session was transformational. The room was full—about 75 men showed up. The leaders who had organized it were surprised by the response. Nobody knew what to expect.

One man spoke up about the pressure to hide emotions, to be strong at all times, and how showing emotion would make him weak. He shared a traumatic incident where he wanted to cry but his mother told him he needed to be strong. Within minutes, other men were sharing similar stories. Fathers who told them tears were shameful. Bosses who dismissed their stress. Wives who expected them to carry burdens silently.

Within the hour, about a third of them were crying, most for the first time in years.

The attendance remained consistently high because we'd created permission to be human. Men told us afterward that it gave them a safe place to talk about things they had never been allowed to talk about before. Not at home. Not with friends. Nowhere.

Men Ki Baat didn't visibly change how people showed up at work. It wasn't designed to. It was designed to give people a place where the mask could come off, where the expectation to be invulnerable could be set down, even if just for an hour. That was enough.

The Resistance Test

At first, our India team was sometimes viewed as an extension rather than a fully integrated part of the company—something I've seen across many organizations expanding globally. Despite repeated conversations, site visits, and clear direction, some struggled to shift from old models of hierarchy to a truly global mindset.

I find this very common in companies struggling to create a global workforce by expanding to India. It's important to understand that the biggest change needs to come from the original team, typically North America. Meeting times, ways of working, communication patterns, relationship building—they all need to be inspected and changed. Without change there is friction.

The resistance showed up in predictable ways.

One leader said, "I don't need to go to India. I've worked with this model for years," assuming that working with any India-based teams was the same. Another told me flatly, "I wasn't hired to work in the evenings. They'll have to flex their hours." Still another suggested, "We'll create recordings of our meetings and they can watch them the next day"—as if asynchronous participation was the same as being in the room where decisions were made.

When resistance came to me directly, I addressed it immediately. I'd step back and talk about the why. "This isn't about India. This is about whether we're building one team or two. If you won't flex your hours, you're saying their time matters less than yours. If you won't go see the operation in person, you're saying you don't need to understand the people you're depending on. That's not the leader we need you to be."

When resistance showed up in group settings, I addressed it there too. I wanted everyone to hear the principle, not just the person who'd violated it. I also coached leaders to do the same—to name the behavior when they saw it and connect it back to our values.

We discovered how easily proximity bias can show up when new teams are formed across borders. Building a shared sense of ownership required awareness and practice. We spent time on the ground together listening, learning, and challenging assumptions. The leaders who embraced this way of working saw the possibility of an evolved way of working cross borders. Others held on to legacy beliefs about where the "real work" happened. That resistance slowed progress and undermined trust.

As with any transformation, not everyone adjusted at the same pace. We had to balance empathy for individuals with the collective need to evolve our ways of working. But empathy had limits. One leader refused to change even after direct conversations about what was required. He left. I didn't force him out, but I also didn't compromise on what the role demanded. Eventually, the misalignment became clear to both of us.

The Shift

But the biggest change happened when individuals went to India and saw it with their own eyes.

One leader came back from Bangalore and told me, "Working hands-on with that team in India just took six months off of the project timeline. We got alignment super quickly and didn't need to schedule any more 'knowledge sharing' sessions. We literally are delivering six months sooner."

Stories like that came back and echoed through the halls.

Leaders who had resisted the trip started asking when they could go. Teams that had been cordial but distant became genuinely collaborative. The narrative shifted from "we have to work with people in a different time zone" to "these are our team members who helped us win faster."

Soon North Americans saw what we had seen from the beginning. First, they encountered incredible technologists—people who lived in a country without any lululemon stores, but who understood the business more deeply than many living right next door to one. They saw engineers who were curious and hungry, who loved the company, and who asked the kinds of questions that prevented costly mistakes. These weren't offshore resources executing tickets. These were business partners who cared about outcomes as much as anyone in Seattle or Vancouver.

Second, they discovered the culture was identical. Walking into the Bangalore office felt like walking into Seattle, Vancouver, or Shanghai. Same clearings at the start of meetings. Same vulnerable leadership conversations. Same commitment to psychological safety. Same Sweatlife mentality—and the India team took it to another level. Eighty percent of the Bangalore tech hub employees participated in the TCS World 10K Marathon, with 90% of them running a marathon for the first time. They trained together for ten weeks, then conquered the finish line as a team. That's what Sweatlife actually looked like when you built it authentically rather than performed it.

Third, they saw the pace. The India team moved fast. Really fast. Without the baggage of "this is how we've always done it," they questioned assumptions, simplified processes, and shipped code at a speed that made our more established teams rethink their own approaches. They weren't constrained by legacy thinking because they weren't carrying legacy anything.

Fourth, they experienced genuine partnership. The India team didn't wait to be told what to do. They identified problems, proposed solutions, and often had working prototypes before anyone asked for them. They took ownership in a way that reminded everyone what "one team" actually meant.

The resistance didn't disappear overnight. But it became increasingly untenable to hold onto old assumptions when the evidence kept proving them wrong.

Building Excellence Through Innovation

Seshan led from within, understanding both lululemon's values and India's culture. Rather than trying to recreate

Seattle in Bangalore, we built something new that honored both. The approach worked spectacularly.

Women, we discovered, were not only more loyal in India, but they became our greatest recruitment tool. Our biggest source of new hires wasn't job boards or recruiters. It was women referring other women. By targeting gender diversity from day one rather than trying to catch up later, we'd unlocked a compounding advantage our competitors couldn't match. But even beyond the recruiting benefit, they became role models. They advocated for themselves and for others.

Our benefits package reflected our commitment to inclusion in ways that went far beyond industry standards. We offered support for IVF treatments, gender transition procedures, and partner benefits for LGBTQ+ employees—even though same-sex marriage wasn't legally recognized in India. We weren't just following local customs; we were taking a leadership position on inclusion because we believed it was both right and smart business.

Our diversity ambitions weren't quotas we grudgingly met, they became a competitive advantage that attracted top talent who wanted to work somewhere that valued what they brought to the table.

The culture that emerged was vibrant and authentically inclusive. We celebrated multiple religious holidays, including Eid alongside traditional Hindu festivals. There was dancing, spontaneous celebrations, and once the team surprised me with a flash mob right in the middle of the office. It wasn't corporate fun mandated from above, it was genuine joy and connection.

The quality of work was exceptional. Innovation was the core DNA of the team. The 24x7 follow-the-sun model we'd envisioned became reality, dramatically accelerating our development cycles. Projects that used to take months were completed in weeks. We were building capabilities that our competitors couldn't match.

This wasn't a cost arbitrage play. This was innovation arbitrage. We were accessing capabilities that would have taken us years to build in traditional markets. These were delivered by people who understood both the technology and our business deeply enough to innovate at the intersection.

We won the Indian trade association NASSCOM's AI Game Changer Award for the merchandise forecasting engine—AI that predicts which products customers will want, when, and where. It was one of retail's hardest problems, and our Bangalore team cracked it.

The award recognized that this was more than a technical achievement. It showed that global innovation at scale is possible.

The Day I Danced

For four years, every time I visited Bangalore, there was dancing. It was part of the DNA. Every celebration, every gathering—the team would form circles, play music, and dance together.

It was vibrant, beautiful, and free. It was part of a cultural memory I didn't share. My family's only music was Sunday mornings sitting in a pew.

They even did a flash mob in my honor one visit. I was so touched…and so glad they didn't invite me in.

I'd smile, clap along, stay safely at the edge of the circle. I'd record videos for my kids, take photos for the team, participate by witnessing. But I wouldn't dance, even when they politely invited me time after time.

Until the day I did not opt out.

On what would be my last trip, the team organized a beautiful celebration to honor our journey together. When they invited me to dance, it felt different. This time it was not expectation. It was love. It was family.

So I danced.

Verna Myers said it best, "Diversity is being invited to the party. Inclusion is being asked to dance." For four years, I'd been invited. That day, I finally said yes.

The bar was alive with circles of people, arms lifted, feet pounding. I moved from one circle to another, following their rhythm. I was not thinking about how I looked or worrying about the phones recording. I was laughing, moving, letting them lead me. In that moment, I felt something deeper than celebration. I felt belonging.

That day, dancing badly but fully, I learned something about leadership I had not understood before. Sometimes the most powerful thing you can do is allow others to lead you into joy. Sometimes the greatest gift you give a team is not perfection, but your willingness to be imperfect with them. Trust is built not only in flawless execution but also in shared vulnerability.

The work we did together was extraordinary. Innovation was our DNA. But it was moments like this, when hierarchy dissolved into human connection, that laid the foundation for everything else we achieved.

For Your Leadership

Building the India Tech Hub taught me that developing human potential creates capability you can't buy or outsource. The team didn't need me to keep pushing them. They were already leading themselves.

See Potential Before It's Proven. Ali Tarhouni taught me he'd hand-picked me for a program years earlier, believing in potential I hadn't demonstrated yet. That became my blueprint for India. Not hiring for credentials alone, but seeing the person underneath. Not waiting for people to prove themselves, but believing they were extraordinary before they'd shown you. The leaders who see potential first build teams that surprise them.

Inclusion Is Strategy, Not Program. Building for 50 percent women engineers wasn't about fairness. It was about unlocking capability competitors couldn't reach. Diverse teams catch bias earlier, ask better questions, design for broader use cases. Homogeneous teams build their assumptions into everything without realizing it. We won the NASSCOM AI Game Changer Award not despite our inclusion strategy but because of it.

Learn Before You Lead in New Contexts. I walked the streets of Bangalore, sat in team members' homes, learned from local leaders before making decisions. The leaders who learn before they lead build something that works. The ones who arrive with answers build something that doesn't survive contact with reality.

Let Others Lead You. For four years, I watched the team dance and stayed at the edge of the circle. On my last trip, I finally joined. Sometimes the most powerful thing you can do is allow others to lead you into joy. Trust is built not only in flawless execution but in shared vulnerability.

CHAPTER 11

The Scale Challenge

The conference room in Vancouver was full of faces on screens. Shanghai. Bangalore. Hong Kong. Seattle. We were one team. But you could feel the center that worked everywhere but felt local—different accents layered over each other in problem-solving sessions, someone in Hong Kong jumping in to finish a thought started in Seattle, the Bangalore team sharing their screens at 9 PM their time because they wanted to be part of the launch discussion.

It wasn't seamless. Working across time zones meant someone was always sacrificing sleep, always stretching their day to overlap with teammates halfway around the world. It meant morning standups at midnight for some, evening strategy sessions at dawn for others. It meant being deliberate about documentation because you couldn't just tap someone on the shoulder. It meant learning to communicate with crystal clarity because misunderstandings compounded across time zones and you might not get to clarify for twelve hours. It meant trusting people you couldn't see, betting on their judgment when you were asleep.

But when it worked—and increasingly, it did work—the complexity became our advantage. Code commits came in around the clock. Teams channels never went quiet. Someone in Shanghai would flag an issue at the end of their day, and by the time they woke up, engineers in Seattle had already pushed a fix and Bangalore had tested it. We weren't handing off work across time zones. We were building together, as one organism that never slept. Someone in Bangalore simultaneously flagged regulatory issues in the EU that nobody in North America had even considered. Someone in Hong Kong caught a performance issue before it hit production because they were testing while Seattle was asleep.

The time zones didn't disappear. They never got easier. But we stopped seeing them as obstacles and started seeing them as infrastructure. Follow-the-sun development wasn't just about speed. It was about having eyes on the system at all hours, perspectives from every region, and the humility to know that the person joining at 2 AM their time might be the one who saves you from a critical mistake.

Five years earlier, I'd been fighting to get budget for a single engineer. Now I was leading conversations about expanding into markets I'd never even visited, discussing international expansion across a dozen countries simultaneously, evaluating acquisition targets that could add hundreds of millions in revenue, planning technology architecture for businesses that didn't yet exist.

The complexity was exhilarating and humbling in equal measure.

How do we maintain innovation velocity when we're no longer the scrappy challenger? How do we defend market position when every startup studies your playbook? What happens when your culture of possibility meets the reality of quarterly expectations?

The answer, it turned out, was that we couldn't—not yet. Not without paying for shortcuts we'd taken on the way up.

When Growth Outruns Its Base

We made deliberate choices to favor speed-to-revenue over long-term infrastructure. One example was product data. Different systems held different attributes about our products, and there was no single source of truth. The website might show one product description, the warehouse system had different specifications, the planning system tracked different attributes, and store systems had yet another version. Every year I considered investing in a unified product data foundation—one master system that every other system would pull from. And every year, I chose not to. Building a unified product data model was a multi-year, multi-million-dollar investment with no immediate return. Meanwhile, new product launches, improved search functionality, or faster checkout experiences could show results in months and drive measurable sales. I made the call to keep moving fast.

And it worked. We scaled. We delivered. We hit our revenue targets. As the company scaled, those cracks became harder to ignore. Personalization, omnichannel, and AI, all of it, depended on product data that didn't exist in a usable form. AI doesn't hide weak foundations—it exposes them faster. Scale without refactoring is just faster fragility.

In the moment, it was frustrating. But we learned that at scale, some foundations don't show their value until it's too late to ignore them. You can defer them for a while, but eventually they become the bottleneck for everything else. The art of leadership is knowing which investments must be bulletproof now and which can wait.

Two years later, I was proven right about the cost of deferring that work. By 2024, the product data mess I'd accepted as a trade-off became impossible to ignore—not because anyone suddenly cared about data architecture, but because AI made the consequences visible to everyone. I had five systems holding the same attributes, color codes that meant different things in different databases, no single source of truth. One retail CEO told me their AI recommended stocking winter coats in Miami because the algorithm couldn't tell "available in navy" from "available in blue." The AI worked perfectly. The data was garbage. That was on me. I'd made the bet that we could outrun the technical debt. For years, we did. But AI changed the speed at which that debt came due.

Every AI vendor promised forecasting, insights, and recommendations—but all of it depended on clean, unified data—something we did not have yet. In one system "navy" meant navy, in another it meant "nightfall." Merchandising categories didn't match fulfillment logic. When humans were making the calls, we could work around the mess. Algorithms can't. They just amplify it.

The lesson was brutal but clear. Technology alone doesn't solve organizational problems. Rather, it exposes them. The competitive take on aren't the ones with the biggest budgets. They're the ones that invested early in the unglamorous work—building foundations strong enough to support whatever comes next.

Our global scaling had worked because it was intentional. We built connection into it—shared time zones, travel, and teams who knew each other beyond the screen. But when remote work expanded overnight, that same pace turned sweatlife into a trap. The foundations that held seams together began to loosen across bedrooms and kitchen tables. What had worked when we saw each other daily in one place started to fracture.

By 2022, the demand for new capabilities was relentless. The business was expanding into new countries while competitors were gaining ground. Guests across the world expected to buy online and pick up in store, return in-store what they bought online, see real-time inventory across all channels. We needed personalized recommendations, better mobile experiences, faster checkout. Supply chains were still recovering from COVID disruptions, and we were racing to build systems that could handle the complexity of global operations.

Every initiative created cascading technology requirements. Launch in a new country? That's payments infrastructure, tax systems, localized experiences, compliance
requirements, supply chain routing. Improve the guest
experience? That touches checkout, inventory systems,
order management, mobile apps, in-store technology.

Roadmaps that looked solid one quarter had to bend
to capture unexpected opportunities or respond to competitive threats we couldn't have predicted.

Global Scale, Human Core

By 2024, the India Tech Hub was no longer new. What
we had opened three years earlier was now fully alive.
The team in Bangalore wasn't just building technology. They were shaping how we worked across borders.
Their approach to innovation, learning, and community
became a model for the rest of the company.

The question became: could we do it again? Could
we scale what worked in India to other regions without
losing what made it work?

China had laid the foundation. We launched there in
2019—two years before India. But China was different.
It wasn't just a tech hub—it was a full market business
with stores, distribution, merchandising, and technology
all embedded together. China proved we could deliver
business results in a complex international market. But
we were still operating in a "get it done" mode, not yet
thinking about the kind of team we wanted to build
while we built it.

India, which opened in 2021, became the proof that the same cultural foundation could work in a standalone tech center. If China showed we could scale operationally, India showed we could scale with intention—building culture and inclusion from day one, not as an afterthought.

India taught us how to build inclusion intentionally, and then we brought those lessons back to China.

In 2019, I invited the women in Asia to a conversation when I visited Shanghai. There were only two in China and one in Korea. Three women across an entire region. I was curious and wanted to know the unique barriers they faced, what it was like working as a woman in tech. Except I failed to recognize that we weren't ready for the conversation. They were reserved—a combination of cultural norms around hierarchy and the fact that they were so outnumbered. The dynamic felt careful, formal, distant. I was asking them to be vulnerable in an environment where vulnerability wasn't yet safe.

What changed wasn't just hiring more women. It was building a workforce intentionally. We brought the men into the conversation. We talked openly about barriers women faced, about unconscious bias, about what it meant to create environments where everyone could contribute fully. We weren't lecturing the men about what they were doing wrong. We were building empathy, helping them see dynamics they hadn't noticed before.

The shift was remarkable. At every Women in Tech meeting in Shanghai after that, the men always wanted to attend. Not because we required it, but because they were genuinely curious. They wanted to learn. They asked questions. They shared what they were seeing in their own teams. And most importantly, they saw the results—these powerful women were driving innovation, solving problems, building capabilities that made everyone better.

My last visit in 2025 showed just how far we'd come. One of my key leaders had built a team of mostly women. We went out to dinner and the change was remarkable. They laughed, asked me tough questions, held their manager accountable when he tried to spin something. They challenged each other's thinking openly. They weren't performing politeness—they were being themselves. Cindy went on that trip with me. They welcomed her with open arms. At the dinner, one of the women asked about our relationship. No judgement, just inclusion.

That transformation didn't happen by accident. It happened because we took what we'd learned in India—about removing barriers, building psychological safety, creating space for people to show up authentically—and we applied those same principles in Shanghai. Different culture, different expression, same foundation.

We learned that global scale meant translating culture, not cloning it. The values were universal. The expression was local.

In India, the team connected through dance and yoga. In China, through walks and Tai Chi. Different forms. Same belonging. Not symbolic acts. Real practices rooted in place.

We also saw innovation begin to flow back from regional teams. R&D experiments in China that reimagined fitting room experiences. In-store analytics that revealed how guests actually moved. In India, architectural patterns and AI models that solved problems we didn't even know we had. Ideas headquarters would never have invented on its own.

My takeaway was simple. Authenticity scales. Uniformity does not.

Principles stay steady. Expression adapts. When people can lead in ways that fit their reality, great ideas travel farther than any mandate.

That is how culture becomes global without losing its soul.

The Leadership Laboratory

The most satisfying part of those final years was watching leaders I'd hired solve problems I'd never even thought of.

Take our Shanghai team. They faced an integration challenge that would have stumped us in the early days—connecting our global inventory system with local payment platforms while maintaining real-time accuracy. Instead of escalating to me or waiting for a consultant, they applied the same systematic thinking we'd developed during our early transformation work. They broke down the problem, identified the constraints, and built a solution that actually improved our global architecture.

In Bangalore, I watched a team combine deep technical innovation with genuine empathy for customer experience in ways that amazed me. They weren't just following the playbook we'd written about putting people first. They were writing new chapters.

The directors who used to operate in silos were now running multi-hundred-million-dollar global platforms. The framework of truth-telling, ownership, and sustainable solutions had become second nature. But more than that, they were improving on it. Finding new applications. Discovering better ways to live out the principles.

I'd call into a leadership meeting in Vancouver and hear someone facilitating a conversation about failure with the same vulnerability and directness I'd tried to model years earlier. But they were doing it better than I ever had, because they'd made it their own.

That's when I knew we'd succeeded. The culture wasn't dependent on me anymore. It was self-sustaining and self-improving. People weren't just following principles we'd established—they were evolving them, making them stronger, adapting them to challenges we'd never imagined.

The transformation had created something bigger than any individual leader or strategy. It had created leaders who could transform whatever came next.

The Foundation was Tested…And It Held

The years that followed tested everything from supply chain breakdowns, to new competitors, to economic pressure, to the pace of AI. But the foundation held.

Not because everything worked. It didn't. Systems failed. Initiatives stumbled. Yet the organization could recover fast because people were honest about what wasn't working and owned the fixes.

Watching leaders I'd hired solve problems I'd never imagined was the moment I knew we'd succeeded. The culture had become self-sustaining. People weren't following my playbook anymore; they were writing their own. That's the privilege of building—knowing what you've created will outgrow you.

Sustainable change isn't about reaching a finish line. It's about building the capacity to adapt to whatever comes next. The platforms could scale. The teams could flex. The culture could stretch without breaking.

Most of all, the people had learned to transform themselves—to stay curious, courageous, and connected, no matter how the world shifted. That's the real return on transformation—creating the leaders for tomorrow's problems.

For Your Leadership

What got you here won't get you there

Systems and practices that were fine at small scale become barriers at large scale. The pilot that works with

100 users breaks at 10,000. You can't wait for failure to evolve your operating model.

Strength test your foundations before scale tests them for you
Data quality, governance, and vendor choices might hold up in early experiments. At enterprise scale, weak foundations crack fast. Fixing it later is slower, costlier, and messier.

Leaders are the real scaling engine
Technology multiplies when people know how to use it well. Develop leaders who understand new capabilities and teach them to others. That is how capability spreads.

Know when to rebuild, not optimize
There comes a point when tuning old systems is wasted time. You have to pause, rethink the architecture, and build the structure that future growth demands.

Success is what continues after you leave
The goal is not to be indispensable. The goal is to make the organization stronger without you in the room. Capability that depends on a single leader is not real capability.

The challenges of scale taught me that even the best strategies have expiration dates. But some things don't scale—they deepen. While I was learning to lead transformation at the enterprise level, my family was teaching me that the most profound growth happens when you let complexity expand your capacity for love.

Part 3:
Leading Forward

The lessons in this section aren't about becoming a more effective executive. They're about becoming a leader whose impact extends beyond your tenure, beyond your company, beyond the immediate transformation at hand.

They're about the humans you develop. The lives you touch. The potential you unlock in people who couldn't see it in themselves.

After two decades of transformation, I've learned that you can build the most sophisticated systems, deploy the most advanced AI, and scale to billions in revenue—and none of it matters if you haven't developed the humans along the way.

What follows is what endures. Not frameworks or methodologies. The lasting impact you create through the people you develop and the courage to bet on yourself when it matters most.

CHAPTER 12

Love as a Leadership Practice

While I was building a culture of psychological safety and authentic connection at lululemon, my personal life was teaching me the same lessons in a completely different context.

When Ermias came home in 2013, I thought I knew his story. I was wrong. What I didn't know would teach me more about leadership than any business transformation ever could.

The story wasn't finished. It was just beginning.

The Rest of the Story

Two years after Ermias came home, when he was eight years old, a friend was heading to Ethiopia on a business trip. I asked her if she would connect with the family who had raised him. I hired the guide who had first taken me there and put them in contact. It was important to us that, even though Ermias didn't want it at the time, he could have a relationship with his Ethiopian family if that became possible someday.

My friend video called me with news from their home. She had found them.

And, Ermias had a six-month-old baby brother named Natan. A brother!

The mother was still inconsistent and left Natan with Wude and Eyayu to raise. I learned they were Ermias' aunt and uncle.

At eight years old, Ermias was all-in with us as an American family, and Ethiopia was a distant memory. He didn't understand what this discovery meant and didn't seem to care too much. He already had a brother and a sister whom he saw every day. This new one in a far-away country he didn't know felt abstract.

My friend returned with a letter, written in Amharic, from the family. I eventually had it translated but immediately recognized the word "Facebook" and the name Bruktawit. I immediately connected with her.

Building Bridges Across Distance

Her cellular connection was very spotty at that time, no Wi-Fi yet in the home, and her English was at a beginner's level. Finding any common words was a challenge. But we kept trying.

Eventually we got a video connection. She had invested every free moment into watching American TV shows to learn English, and it worked. I am still not quite sure how she did it, but one day when I opened the video chat, she was speaking English. Stubborn love crosses borders, evidently. We began to talk regularly.

At first, Ermias had no interest. His life was good and his memories of Ethiopia were not. It was difficult to explain to this family why he wouldn't show his face on video. I explained to them that he was not ready, and he needed to control the timeline. Both Bruktawit and I persevered because we shared the love of Ermias. She was his cousin and was raising Natan, Ermias's younger brother.

The birth mother of the boys had both mental and physical concerns that prevented her from staying in their lives. They asked if we would take Natan to live with Ermias. Adoptions from Ethiopia were closed at that time, but we provided financial support so they could keep him, send him to a private school, and give him opportunities to thrive with the only family he knew.

As our relationship with Wude, Eyayu, Bruktawit, and their family deepened through weekly video calls and daily texts, I faced a decision that would test everything I believed about authenticity and relationship.

The Truth I Couldn't Keep Hiding

During the adoption process, I had presented myself as a single mother to the Ethiopian government. Cindy was described as a friend and caregiver. Being gay was illegal, and since we had never had a formal ceremony, we were able to walk a fine line. It was the only way to ensure Ermias could come home to us.

But as we developed a relationship with the family there, it didn't feel right. Cindy was his mother and was being introduced as an aide.

I could no longer live this lie. I had done it too many times before in corporate settings to know how it limited real connection. The stakes felt enormous. This was

Ermias's family. In a country where 97 percent of people believe homosexuality should be rejected by society, I was risking the most important relationship in my son's life.

I consulted an Ethiopian friend living in the United States. He spoke in Amharic and told Bruktawit the truth about Cindy and me.

Her reaction? She was completely nonplussed. Matter-of-fact acceptance. Just love. We were family and that mattered more than anything else.

When she shared it with the rest of the family, nothing changed except we grew closer and the wanted to know Cindy more. This was their first time experiencing a gay person. Didn't matter. Everything felt pure and genuine. Nothing but family and love.

What Their Acceptance Taught Me

Their acceptance taught me something profound about the difference between institutions and individuals, between cultural norms and personal relationships. While Ethiopians society put law might reject who I was, this family—Ermias's family—simply saw love. They saw commitment. They saw two people raising their beloved boy together.

From that experience I learned that when you create space for people to show up as fully human, barriers that seem insurmountable become secondary to connection. It had been so focused on what the culture said was impossible that I nearly missed what the individuals were ready to offer.

Years later, that lesson would show up in unexpected places, such as a virtual call during wartime.

Russia and Ukraine, 2022

Early in the Russia-Ukraine war, I had team members from both countries. Emotions were raw. Teams didn't know how to interact with each other. So I worked with our People & Culture partner to host a listening session. We set ground rules to create a safe space. Take space, make space, hold confidence, no judgment, respect, curiosity over assumptions.

I opened with one question, "How are you doing?"

Then I waited.

Hundreds of people on the call. Dead silence. One minute, maybe two. I kept waiting.

Finally, someone spoke. Ukrainians first. Fear for their families' lives. Loss of their country. Terror of what was coming. I had a pit in my stomach. What was I thinking? There was nothing I could solve. They were at war. Two sides, no middle ground.

Then a Russian man began to speak. We all held our breath.

"You are our brothers and sisters," he said. "We grew up beside you." He explained the pressure he was feeling in the United States. How his last name made him identifiable. The judgment that came with it.

What happened next still gets me. Each side took a step toward each other. Tears started, hundreds of people on that call, emotions running higher than I'd ever seen.

But our culture held. Everyone extended trust. Respect. Deep care for their teammates.

That moment taught me that listening, deep listening, opens pathways to trust. Humans seeing humans is stronger than fear.

And it doesn't mean we lowered our standards or softened our edge. In fact, the opposite happened. When people feel genuinely seen and cared for, they bring more—more creativity, more resilience, more willingness to push through hard things together. The teams that extended trust to each other during that call became some of our highest performers. Care didn't make us less competitive. It made us unstoppable.

That same principle—that deep care across distance creates strength, not weakness—was proving true in my personal life too.

Becoming Family Across Continents

We first met as a family in Tanzania in 2021. All our kids—Kenzie, Mason, Ermias and Natan—developed sibling bonds immediately. On safari, we watched Ermias and Natan teach each other words. Natan only spoke Amharic. Ermias had stopped speaking it years earlier, but with Natan, fragments came back. "Elephant" in Amharic. "Lion" in English. They'd point, laugh, correct each other's pronunciation. Two boys finding a language between them that was part Amharic, part English, part sibling.

We asked Wude and Eyayu if we could sponsor Natan in an international school so he'd have access to stronger education. They said yes. Today, Natan is a spicy, loving, funny 10-year-old who speaks near-perfect English. We just put his tenth birthday presents in our suitcase before a family trip to the Philippines, where we all celebrated together.

We have a global family now. We love across borders and time zones. We video-chat every week, text every day. Distance doesn't diminish connection when there's genuine commitment. With intention and investment, you can maintain and deepen relationships across any distance.

More importantly, expansion doesn't require replacement. Ermias doesn't have to choose between his Ethiopian family and his American family. He gets to have both. That understanding became fundamental to how I approached global teams. People don't have to choose between their local culture and our company culture. They get to have both. The best teams integrate rather than assimilate.

What Ethiopia Continues to Teach

Every major leadership decision I made after Ethiopia was informed by what I learned there. When I chose internal teams over consultants at lululemon, I was applying the lesson that ownership matters more than polish. When I pushed to hire for a balanced male/female workforce in our India team, I was remembering how much talent gets overlooked when you only hire from familiar patterns. When I banned "onshore/offshore" language, I was thinking about Ermias and the power of words to include or exclude.

The foundation of my leadership transformation wasn't just technical or strategic. It was deeply personal. It was learning to see strength I hadn't recognized, to value contributions I hadn't understood, to build belonging that didn't require people to lose themselves in the process.

People sometimes ask what adoption taught me about leadership. The answer is everything. The most important transformations happen not when you change people, but when you let people change you. Real strength often looks different than what gets celebrated in boardrooms. The best teams, like the best families, aren't built on similarity, but on commitment to each other's growth across difference.

Ermias didn't just join our family. He expanded what family could mean. And that expansion—of perspective, of possibility, of what it looks like to truly belong to each other—became the foundation for every team I would ever build.

While I was helping scaling lululemon from \$2B to \$10B, building teams across continents, navigating crisis after crisis, my Ethiopian family was teaching me that the principles I was applying at work weren't just business strategy. They were how humans actually connect across differences.

Authenticity. Psychological safety. Integration over replacement. Letting people define belonging for themselves. Being changed, not just changing others.

These weren't leadership buzzwords. They were how I built my family. And that's why they worked when I built my teams.

The lessons from Ethiopia would prove essential in what came next. Because the real test of influence isn't what you can do when everything is going well. It's what endures when you're no longer in the room.

My greatest satisfaction no longer came from being needed, but from watching others realize they didn't need me anymore. It was time to ask a question I'd never seriously considered: what if the transformation was complete?

CHAPTER 13

Onward

The Question That Changed The Answer

Five years ago, I was at an event for women leaders in technology at lululemon. We were talking about goals, which is what I've always lived by. The facilitator went around the room asking each of us what we were working toward.

When she got to me, I froze.

For the first time in my career, I didn't have an answer. I'd spent decades chasing the next thing, the next promotion, the next transformation, the CIO role I'd worked toward for years. But sitting in that room, having achieved what I'd set out to achieve, I realized I had no goal beyond where I was. I was happy. The work mattered. The team was thriving. But I wasn't working toward anything anymore. I was just... working.

"Julie," the woman asked again, "what goal are you working toward?"

I said the first thing that came to mind, "to write a book," and then had a look of surprise that made the room laugh after I said it.

I hadn't planned to say it. A few people had, over the years, mentioned that I should write a book, but I hadn't been thinking about it. However the moment I said it out loud, something shifted. It felt true in a way I hadn't expected.

The thing about getting a bunch of powerful women together and giving them your goal is they hold you accountable. Over the next months, they kept asking: "How's the book coming?" "Have you started writing?" "When can we read it?"

So I started writing. Stories about my dad teaching me systems thinking from behind home plate. About that twenty-hour website outage before I even started. About building the India Tech Hub with humility instead of arrogance. The stories piled up, but they didn't create a compelling book. They were just... stories.

A few months later, I mentioned the book Calvin, lululemon's CEO. I wasn't asking for anything. Just sharing something I was working on. He made it clear that the time wasn't right to write it from within the company. And I knew that book that would tell the truth about leadership, transformation, and what it really takes to build inclusive culture would require time and space I didn't have while running technology for a global retailer scaling at breakneck speed.

That conversation planted something I couldn't shake: What if my great job was preventing me from growing? Had I become complacent?

The Uncomfortable Truth

I didn't want to think about leaving lululemon. The work still energized me. The team was the best I'd ever built.

We were scaling innovation across continents, proving that culture could be a competitive advantage, building something that would outlast all of us.

But I'd opened a door I couldn't close. The book would force me to articulate what it actually means to create impact that reaches beyond the walls of any single company.

For years, people called me Chief Information Officer. But the work that mattered most was never about information. It was about impact, and influencing humans who could transform organizations long after I was gone, creating cultures where truth-telling was possible and psychological safety enabled risk-taking, developing leaders who understood that technology amplifies human capability rather than replacing it.

Chief Impact Officer. That's what I'd actually been building toward.

The impact that lasts isn't the systems you build. It's the humans you develop who keep building after you're gone. The engineer who learns to challenge assumptions. The manager who discovers they can lead without authority. The team that moves from fragmented to unified because someone showed them what trust makes possible.

That impact doesn't scale by staying in one role at one company. It scales by helping leaders everywhere build these capabilities intentionally. Through boards where I could influence multiple companies. Through advising leaders facing transformation they'd never navigated before. Through writing and speaking where the lessons could reach people I'd never meet.

The impact I wanted wasn't just organizational. It was beyond organizational. And staying at lululemon, no matter how much I loved it, would keep that impact contained.

Every time I sat in a leadership meeting, I found myself watching the team with different eyes. They didn't need me the way they used to. They were solving problems I hadn't thought of. Making decisions I would have made, or better ones I wouldn't have considered.

The Korea launch that taught me about earning trust across distance? My team was now doing that in six markets simultaneously without my involvement. The product data infrastructure challenges that had exposed themselves through AI? We had finally prioritized the foundation work and were rebuilding it properly. The psychological safety we'd spent years building? It was operating without me reinforcing it in every meeting.

The transformation I'd led had succeeded in the most important way. It no longer needed me to drive it.

That realization scared me more than I wanted to admit.

For thirty years, my identity had been built on being needed. On being the person who could see the whole field, solve the complex problem, lead through the crisis. What would I be if I wasn't that person anymore?

The Wrestling

I talked to Cindy about it late one night, after the kids were in bed. "I think it might be time to leave," I said. Saying it out loud made it real in a way that thinking about it hadn't.

She looked at me carefully. "Are you leaving because something's wrong, or because something's right?"

That question cut through everything. Nothing was wrong. The team was thriving. The culture was holding. The systems we'd built were scaling. I wasn't running from failure. I was running toward something I couldn't quite name yet: the possibility that my impact could be bigger outside lululemon than inside it.

The Decision

What if staying was actually holding me back? I carried that question for two years.

I didn't rush the decision. I sat with it. Wrestled with it. Watched the team continue to grow and thrive. Every leadership meeting reinforced what I already knew: they were ready. The transformation had worked. The culture was self-sustaining. They didn't need me the way they used to.

But leaving still felt enormous. This wasn't just a job. It was eight years of building, transforming, proving that culture could be infrastructure and psychological safety could drive speed.

I didn't rehearse for the meeting. I don't remember what I thought about on the way there. I just remember saying it out loud, "I think it's time for me to go."

Calvin didn't look surprised. Maybe he'd seen it coming. Maybe leaders always see when other leaders are ready for what's next. He asked me to stay through a successful transition, which I gladly did. Not because I had to, but because leaving well mattered as much as anything I'd built.

The Evolution of Impact

For most of my career, ambition looked like upward movement. Bigger titles. Larger teams. More responsibility. Proving I belonged in rooms I'd been told weren't meant for people like me.

But somewhere along the way, ambition shifted. It stopped being about what I could achieve and became about what I could help others discover in themselves. The impact I could have.

The moments that mattered most weren't the wins that made headlines. They were the conversations where someone realized they were capable of more than they'd imagined. The teams that solved problems I couldn't have solved for them. The leaders who became better than I'd ever be.

That's when I understood that the goal wasn't to be indispensable. It was to make yourself unnecessary.

When Leadership Becomes Service

There's a version of leadership that's about being the smartest person in the room. The one with answers. The hero who saves the day.

I'd spent years trying to be that leader. It was exhausting. And it didn't scale.

The shift happened gradually, then suddenly. I stopped needing to have the answer and started asking better questions. Stopped directing and started creating conditions where teams could figure it out themselves. Stopped measuring my value by how many decisions I made and started measuring it by how many decisions my team could make without me.

Leadership became less about control and more about trust. Less about being right and more about helping others find their own way to right answers.

That shift didn't make me less valuable. It made me irrelevant in the best possible way. The team kept moving without needing me to push them. That's when I knew the transformation was real.

Walking Out The Door

During my last month at lululemon, I felt truly celebrated. My teams showered me with love and appreciation in ways that touched me more deeply than I expected. It wasn't about the list of accomplishments we'd achieved together, the transformation to a $10B company, the India Tech Hub, the pandemic response, the cultural foundation that had become our competitive advantage. It was the notes telling me how I had impacted their lives. The impromptu "can I tell you something" moments, such as someone pulling me aside to share one thing I'd said years ago that they still carried with them. The hugs. Even the tears.

The kind and wonderful Teams messages, texts, emails and even cards. These weren't just nice words. They showed me evidence of multiplication. *Impact*. People I'd influenced were now influencing others. Leaders I'd developed were developing other leaders.

That's when I knew we'd built something real. Not because of what we'd accomplished, but because of who people had become in the process. They weren't thanking me for solutions I'd provided. They were thanking me for permission I'd given—to be authentic, to lead with vulnerability, to trust their own judgment.

And, of course, the dancing.

I left lululemon knowing the work would continue without me. That was the point.

What Endures

Six months after I left lululemon, I started getting messages. Not from the team, though those continued too, but from other leaders. CTOs and CIOs who were stuck in the same patterns I'd been stuck in years earlier. Performing competence instead of admitting uncertainty. Hoarding authority instead of distributing it. Building systems that needed them instead of teams that didn't.

They'd read something I'd written or heard me speak and wanted to talk. Not about technology architecture or AI strategy. About how to build cultures where truth-telling beats self-protection. About how to develop leaders who become better than you. About how to know when staying is actually holding everyone back.

I realized that the impact I'd hoped for was happening. Just not the way I'd imagined.

I wasn't building systems anymore. I was helping other leaders see what was possible. The capabilities I'd developed through decades of transformation from navigating power dynamics, to making sense of chaos, to staying grounded when algorithms offered false certainty, to building sustainable velocity, to unlocking potential others overlooked. Those were exactly what other leaders needed as they faced their own AI transformations.

What I took with me wasn't the awards or the revenue growth or the systems we'd built. It was the knowledge that transformation isn't about what you accomplish while you're there. It's about what endures after you're gone.

And what endures isn't systems or processes or technology. It's capability. The ability of people to solve problems you never anticipated. To lead in ways you never taught them. To become more than you could have imagined for them.

The Deeper Discovery

Writing this book forced me to see patterns I'd lived but never fully understood.

The lessons didn't start in boardrooms. They started behind home plate, where my dad taught me to see the whole field. In my mom's upholstery shop, where I learned that strength doesn't announce itself. In the hiding years, where I discovered that masks protect nothing and cost everything. In that coffee house in Addis Ababa, where strangers taught me that connection transcends what we think are barriers.

Every transformation I led was built on those foundations. The twenty-hour website outage before I even started at lululemon? I knew how to respond because I'd learned ownership from watching my mom pull a needle through her thumb and go back to work. The decision to bet on our internal team instead of consultants? That came from understanding that you can't outsource belief. Building the India Tech Hub with gender balance as strategy? I could do that because Ethiopia had taught me to learn before leading.

The capabilities that matter most in leadership, such as seeing systems, standing firm, owning outcomes, developing others, staying authentic under pressure, those don't come from business school. They come from experiences that force you to develop them.

And I'm still learning. You don't master these capabilities and move on. You keep practicing them, keep discovering new edges, keep finding places where you're still performing instead of being real.

The Real Test

Months after I left lululemon, the universe tested everything I believed.

I'd spent those months doing the work. Connecting with CEOs and board members. Learning from founders. Saying yes to conferences, speaking, education. Finishing this book. Rebuilding the child-like curiosity that happens when you don't have anything to fall back on. Discovering what I was capable of when no one was setting my agenda but me.

The doubts were real. The fear was real. I'd had a full-time job since high school. Who was I without that structure, that title, that identity?

Then the call came. The job of a lifetime at an incredible global brand. Everything I had dreamed of in an ideal job: the culture, the history, the future potential. I spent months interviewing, building relationships, imagining myself in that role.

And I walked away from it.

Not because it wasn't amazing. It was. But something unexpected had happened in those months of reinvention: I'd built something that mattered more.

The dream of building something for myself was working. The book was resonating. Board opportunities kept coming. Speaking opportunities appeared. Advisory work was thriving. And this life where I controlled my daily choices, where I was building something entirely my own? I wasn't ready to give that up.

Here's what I learned. Setting goals is important. Doing the work is essential. But the hardest part? Betting on yourself when it counts. When there's a safer option right in front of you. When everyone might think you're crazy for walking away.

For thirty years, I'd built my identity on being needed. On being the person who could see the whole field, solve the complex problem, lead through the crisis.

Leaving lululemon was hard, but I was leaving something I'd completed. Walking away from this opportunity was different. I was choosing uncertainty over security. Choosing the life I was building over the life being offered.

That's the real test of everything in this book. Not whether you can lead transformation inside a company. Whether you can lead transformation in yourself. Whether you can trust the capabilities you've developed enough to bet on them when it matters.

The Intentional Life

A few weeks ago, I was on a call with a CIO wrestling with whether to stay in his role. The company was thriving. His team was strong. But he felt stuck.

"I don't know what's next for me," he said. "I've accomplished what I set out to accomplish. Now what?"

I recognized that feeling immediately. It's the same question that woman asked me five years ago: what goal are you working toward?

"Maybe the question isn't what's next," I said. "Maybe it's what matters now that you can finally see it clearly."

He was quiet for a moment.

"For years, you've been climbing. Proving yourself. Building your career. Transforming the company. All of that mattered. But now you're in a position where you can actually choose what to optimize for. Not what looks good on a resume. Not what impresses the board. What really matters to you."

Another pause. "I think... I think what matters is helping other people figure out what I wish I'd known earlier."

"Then that's your answer."

That conversation stayed with me because it wasn't just his answer. It was mine too.

What gives work meaning isn't the title or the comp or the scope. It's the *impact* you have on people's capability to lead their own transformations. To see what they couldn't see before. To become who they're meant to become.

The future belongs to leaders who can build influence when authority isn't enough. Who can navigate complexity without needing perfect information. Who can stay human when algorithms offer efficiency at the cost of wisdom.

That future is being built right now. By people willing to do the uncomfortable work of developing these capabilities. By leaders who understand that transformation isn't about technology. It's about what technology makes possible when humans are ready for it.

If you're in the middle of your own reinvention, wondering if it's working, if you're enough, keep going. Do the work. And when the moment comes to choose yourself over the expected path, I hope you find the courage to do it.

The intentional life is worth it.

Appendix A: Leadership Summit Transcript

Hi everyone, I'm Julie Averill.

I'm the Chief Technology Officer here at lululemon.

It's an extreme honor and privilege to be here today.

And if I'm being honest, it makes me a little bit uncomfortable, but that's OK.

I have spent more than half my life hiding my true identity, my true self from almost everyone who knew me.

I think people have a hard time understanding and how I waited until my daughter was two years old speaking in full and complete sentences to come out to my parents.

I hid everything that was really important about myself.

My values, my true feelings and the person that I loved the most in the world from everybody except a handful of people.

One thing that I did love and have loved since I was very young was technology.

My dad put me in a training class at RadioShack in the 5th grade and I just developed a very special bond with solving hard problems and connecting with the logic an intensity that I found with computers.

So much so that I don't think I even noticed that I was the only kid in class, let alone the only girl.

But when I was still taking some of those same classes in my computer science studies in college I noticed.

And I would get questions like what's it like to be the only woman here, or what brought you here and those questions made me feel uncomfortable.

They made me feel like I didn't belong and so I didn't necessarily reach out to people I didn't want to answer those questions.

And so I just retreated.

At the same time, I was escalating in my career and I was also noticing that I was becoming the only woman in this predominantly male club and I didn't want to stand out.

I didn't want to answer some of those same questions and really, even to myself of do you really fit in here?

And so I just tried to blend in.

And by that time I also knew that I was gay and that presented its own set of fears and risks, and I just didn't want to talk about that at all.

I thought I could lose my job over it.

I had no workplace protections, and I didn't know how my family would respond,

But there was nothing that was worth taking that chance of not being part of my family,

So I just continued to build this wall that I hid behind.

And the topics of conversation with anybody became very safe.

It wasn't until I got my first CIO job that I reached out to a mentor and I had a lot of questions for him and I thought I would get answers and instead I got one very poignant question back at me and his question was Julie, how are you going to get your new team to follow you?

And I thought that was a very easy question.

I tried to answer it, but he actually knew me pretty well and he said to me, Julie, you have a wall built around you and nobody knows the real Julie.

How are you going to get these people to follow you?

And it was such a piercing question, and I knew that in order to become the leader that I wanted to be, in order to become really the daughter that I wanted to be, the wife that I wanted to be, the mother that I wanted to be, I had to answer that question.

And it was not easy to answer that question and it was not instantaneous and it's been a journey for me.

The journey included looking inside myself and understanding what my true values are, and what I was willing to really stand up for, and what I was willing to risk.

And the walls started coming down and once they started coming down I began to know freedom and I started knowing what it's like to really live in authenticity

And I can tell you today that I stand here as the Chief Technology Officer of this incredible company and I stand before you truly free.

I stand firmly rooted in my beliefs that diversity is a key unlock.

It's not just something that is good for us to do, but it is an unlock for us as a company in our strategic values.

And these principles of inclusion diversity equity and action are things that will take us to a place where we can all live truly free.

And truly free means that we bring other people to the table, and we're enriched by the diversity of everyone around us.

This journey is not easy. But it is worth it, I promise you.

Acknowledgments

This book was never just mine. Every chapter carries the fingerprints of people who helped me learn, fail, grow, and find my voice.

To my parents, Earl and Pat Averill, who gave me the foundation of strategic thinking and quiet strength that anchors everything I do. To my siblings, who taught me that real relationships require moving from performance to presence.

To Cindy, my wife, who saw me clearly when I was still hiding and gave me the space to become myself. You've been my anchor through every transformation, personal and professional. You made space for me to be uncertain, scared, and figuring it out. You never asked me to have all the answers. You just asked me to be real. That's the gift that made everything else possible.

To our children, MacKenzie, Mason, and Ermias, who teach me daily about fearless authenticity and expanding love.

To our Ethiopian family—Wude, Eyayu, Bruktawit, Azeb, Mikeus, Natan, and little Evana Jula—who welcomed us not as outsiders but as family. You taught me that true love transcends all boundaries and showed me lessons about belonging and leading across difference that no framework ever could.

To my friends who modeled authenticity so fearlessly that staying hidden became harder than coming out. Your courage gave me permission to find my own.

To the Women in Tech who asked the question that opened the door. Thank you for your friendship, for pushing me, for teaching me, and for holding me accountable.

To the technology teams at Nordstrom, REI, and lululemon who trusted me to lead them through transformation. Especially to Venki Krishnababu, whose partnership made impossible timelines possible, and to Vaidynathan Seshan, who built our India Tech Hub with a vision of what's possible when we lead with values. To every leader, engineer, architect, and project manager who chose to build the future with courage and care.

To the global lululemon team who welcomed me with such generosity and taught me what authentic cross-cultural leadership looks like.

To Mike Richardson, who saw through my masks before I knew I was wearing them and asked the question that changed everything: "How are you going to get your team to follow you when they don't even know who you are?" You taught me that trust is the real currency of leadership.

To every person who shared their story with me, who trusted me with their truth, who made me a better leader by letting me see their full humanity.

Cindy, Julie, Mason, Ermias, Kenzie, Bruktawit and Natan (in front)

Notes

1. Challapally, A., Pease, C., Raskar, R., & Chari, P. (2025). "The GenAI Divide: State of AI in Business 2025." MIT NANDA Initiative

2. Hunt, Vivian, Sara Prince, Sundiatu Dixon-Fyle, and Lareina Yee. *Delivering Through Diversity: How Inclusion Matters.* January 2018. McKinsey & Company.

3. https://corporate.lululemon.com/media/press-releases/2018/03-27-2018-085857433

4. https://corporate.lululemon.com/~/media/Files/L/Lululemon/our-impact/reporting-and-disclosure/2020-impact-agenda.pdf

5. https://www.outlookindia.com/national/transgender-and-unemployment-in-india-news-182617

6. World Bank, *The Equality Equation: Advancing the Participation of Women and Girls in STEM* (Washington, DC: World Bank, 2020); confirmed by India's All India Survey on Higher Education (AISHE) Report, 2019-20.

7. Akanksha Khullar and Nalini Bhandari, "Over 40% of Indian STEM Graduates are Women – But What About Jobs?" *The Quint*, July 23, 2022 (citing UNESCO data).

8. "Why Marriage Costs Indian Women Their Jobs While Boosting Men's Careers: Marriage Penalty," *Business Standard*, October 16, 2024. Women's employment rates drop by 12 percentage points post-marriage.

9. Ibid. Men's employment rates rise by 13 percentage points after marriage, creating a "marriage premium."

About the Author

As Chief Information Officer of lululemon, Julie Averill led the technology transformation that scaled the company from $2 billion to over $10 billion in revenue while building global teams capable of sustaining that growth. Prior to lululemon, she led omni-channel and digital transformations at Nordstrom and REI, navigating system failures, high-stakes crises, and the complicated work of integrating technology with business strategy at scale.

Julie spent her career navigating systems that weren't designed for her—and helping change them along the way. A female leader in technology. A gay woman in rooms where difference still raises eyebrows. A mother who carried two and adopted one child while leading global technology organizations. And a Chief Information Officer who believes the future of AI depends less on machines and more on how human we're willing to be.

Today, she advises boards, CEOs, and founders at the intersection of AI capability and organizational readiness—the place where most transformations stall. She speaks globally on leadership in the age of AI and helps organizations build what technology can't deliver: teams that tell the truth, leaders who can turn pilots into real change, and cultures that sustain transformation instead of performing it.

After years in corporations proving herself, building teams, and transforming companies, Julie learned that what matters most is not the systems you build or the revenue you generate, but the people whose capabilities you unlock, the leaders who become more than they thought possible, and the ripple effects that continue long after you're gone.

She lives in Bellevue, Washington, with her wife Cindy and their three children. When not speaking, advising, or writing, she's on the pickleball court, traveling to visit family across continents, or learning from her global family that the most profound transformations happen when we expand who we're willing to become.

Other titles from 8080 Books

No Prize for Pessimism by Sam Schillace
A Platform Mindset by Marcus Fontoura
WorkLab by Collette Stallbaumer
Human Agency in a Digital World by Marcus Fontoura
Hyderabad Days by Ravi Vedula
Showstopper! by G. Pascal Zachary
Leading with Imperfect Feet by David Dame

Check for new titles on our website:

https://unlocked.microsoft.com/8080-books/